SpringerBriefs on Pioneers in Science and Practice

Volume 29

W0234477

Series editor

Hans Günter Brauch, Mosbach, Germany

More information about this series at http://www.springer.com/series/10970
http://www.afes-press-books.de/html/SpringerBriefs_PSP.htm
http://afes-press-books.de/html/SpringerBriefs_PSP_Gleditsch.htm

Nils Petter Gleditsch

Nils Petter Gleditsch: Pioneer in the Analysis of War and Peace

Nils Petter Gleditsch
Peace Research Institute Oslo (PRIO)
Oslo
Norway

Acknowledgement: The cover photograph was taken by PRIO. The other photos is this book were taken (in the order they appear in the book) by Eva Koch, an unidentified news photographer, Nils Petter Gleditsch, Aage Storløkken at Scanpix, Arne Pedersen at Dagbladet, and Stein Tønnesson. All pictures are reprinted with permission. A book website with additional information on Nils Petter Gleditsch, including videos and his book covers is at: http://afes-press-books.de/html/SpringerBriefs_PSP_Gleditsch.htm.

ISSN 2194-3125   ISSN 2194-3133   (electronic)
SpringerBriefs on Pioneers in Science and Practice
ISBN 978-3-319-03819-3   ISBN 978-3-319-03820-9   (eBook)
DOI 10.1007/978-3-319-03820-9

Library of Congress Control Number: 2015938731

Springer Cham Heidelberg New York Dordrecht London

Copyediting: PD Dr. Hans Günter Brauch, AFES-PRESS e.V., Mosbach, Germany

Printed on acid-free paper

Springer International Publishing AG Switzerland is part of Springer Science+Business Media
(www.springer.com)

*This book is dedicated
to my former students and to
younger colleagues in peace research
everywhere who are taking
the analysis of war and peace
to new levels that we could only have dreamt
of when I started my own work*

Nils Petter Gleditsch on his way to the Skåla mountain (1843 m) in Loen, Norway to celebrate his 70th birthday in 2012. *Photo* Eva Koch

Foreword

Nils Petter Gleditsch: A Peace Research Pioneer

From an early date, Nils Petter Gleditsch was a well-known person in the Nordic peace research community. He had joined the *Peace Research Institute, Oslo* (PRIO) in 1964. At that time, as my then fiancée Lena (now wife) and I observed when we arrived in 1967: PRIO had all the enthusiasm of a pioneering environment. I did not meet Nils Petter in his own milieu until a year later. In 1967, he was at one of the centres for peace research at the time: University of Michigan in Ann Arbor. Its Center for Conflict Resolution is often mentioned as one of the early originators of peace research in North America (Wiberg 1988). Certainly this is where the *Journal of Conflict Resolution* was born, but by this time it was in crisis. Importantly for Nils Petter Gleditsch and for peace research, this was where J. David Singer had just initiated the seminal Correlates of War project. It was exciting in that it used new approaches (systems analysis), new forms of data collection (punch cards!) and new statistical methodologies.[1]

In Oslo, however, Johan Galtung was the key person, only 36-years old, but already the 'guru' for many. Few of us in the younger generation got to know him as well as Nils Petter did. Galtung originally defined himself as research leader, 'forskningsleder'. The researchers of the institute were against hierarchy, thus PRIO aimed to have a flat structure. I encountered skepticism to academic degrees and academic traditions. To be an independent institute was deemed preferable and gave researchers greater leeway than in a traditional structure. It certainly did not prevent PRIO researchers from getting involved with university departments, and Galtung was appointed to the chair in conflict and peace research that was established at the University of Oslo in 1969. Galtung told us that he had deliberately not pursued a Ph.D. as a reaction against an arrogant and conservative system. Nils Petter shared this sentiment. As testimony to

---

[1]A volume edited by J. David Singer (1968) provided first insights into the possibilities of quantitative measures for peace research.

their brilliance, both these men became professors at universities, without having the normal entrance ticket to the academic life of a Ph.D. degree!

At this time, Nils Petter was engaged in a project dealing with air travel. He was mapping the actual routes of commercial airlines in the world. The project aimed to understand the structure of world interaction with the help of this simple indicator. It may have said more about PRIO than we realized at the time. The project focused on 'interaction' as a key concept. Clearly inspired by recent advances in sociology, notably Homans (1961), it developed the so-called contact (or exchange) hypothesis, which talked about the beneficial impact on peace of more contacts. There was considerable interest in interaction, and my own project at PRIO dealt with the opposite: sanctions, the breaking off of interaction. One could perhaps even talk about 'negative' and 'positive' interaction, in line with the description of concepts of peace in the famous editorial of the first issue of *Journal of Peace Research*.[2] We all had read it, of course, and assumed that it was written by Galtung, although it was not signed.

However, Nils Petter was doing more than the contact hypothesis. He was concerned with the 'structure' of interaction, that is: which parts of the world were tied together and which ones were not. The message became extra powerful, when he showed a map of the world in terms of what a more equitable distribution of world air routes would look like. There would be more connections between South America, Africa, and Asia than were provided at the time. There was, he explained, also a 'rank' element: the centers were interacting more, the peripheries less. It turns out that the air routes were a good indicator of global integration or lack thereof. Nils Petter had found a measurement that is still relevant and the picture can be confirmed by any traveller today; just by pulling out the airline magazine from the seat pocket in front of all air passengers. Even today, the patterns Nils Petter demonstrated remain a testimony to the existence of a solid structure of interaction.

The study appeared in *JPR* (Gleditsch 1967). It served as an empirical validation, or, perhaps even as an inspiration, for Galtung's later works on the structure of imperialism. From the study of interaction it was not difficult to make sense of the politically more loaded term of 'imperialism' (Galtung 1971).[3] Nils Petter certainly saw this, and in a related article (included in this collection), he suggested the notion of 'time imperialism', how those in the center of interaction networks also can control the setting of schedules (cf Chap. 3 of this volume). Even today, we can observe such imperialism by assessing from which spots one can travel the most comfortably in terms of the daily flight schedule and where one has to board the plane in the middle of the night.

What is striking in this story is the role *JPR* already played at this time. First of all, the journal was widely read. It quickly occupied a central position in the peace research community. Every issue was read in detail (at least I speak for myself) to see what new topics, methods, or sources were being used. Critical articles also received a forum in the journal. Second, the journal's policy was to invite contributions from

---

[2]An Editorial, *Journal of Peace Research* 1964, 1(1): 1–4.

[3]This was one of many Galtung articles in this period with the term 'structure' in the title.

young researchers. Nils Petter was 25 when his first article was published, and several other authors were the same age. Few journals have displayed such strong support for younger scholars. Third, it also encouraged new milieus of peace research, by offering them special issues. Certainly a risky approach from a journal's perspective as the contributions may vary in quality and not appear on time. Still, it was very important for aspiring groups to develop credibility in a field that could have been easily marginalized. Fourth, the logical and empirical strength and innovation of the articles drew scholarly attention and were widely quoted.

By the late 1960s, Nils Petter Gleditsch was already an established resource at PRIO. He has remained a backbone of the organization. He personally constitutes its historical memory. He has served in most functions of the institute, even being involved in the controversial policy of equalizing the salaries of all employees that was executed in 1972. He repeatedly served as director of PRIO in the 1970s when directorships were rotated on an annual basis among leading researchers in the institute. This was a way to deal with the departure of Galtung, replacing his rule with a more 'democratic' structure. The inner life of PRIO would constitute a fascinating study of organizational experiments, including what such a test could do to productivity. The present PRIO incorporates some of the features of those trials, although on the surface it may look like any other research organization, the history tells us something different and the person to go to for information is Nils Petter.

Gleditsch's wisdom came to its optimal expression as the editor of *Journal of Peace Research*. *JPR* was central to PRIO from the beginning, but Nils Petter made it crucial not only to peace research but to all social sciences dealing with politically relevant matters of world affairs. This is no small achievement. He took up the editorial position in 1983, although having served for a short spell in the 1970s (another element of the rotation principles that characterized the institute at that time). He held it for the following three decades. It was a remarkable period of the growth of the journal. He established it as a leading, high-quality journal. The commitment of the editor as well as the editorial staff, the many reviewers and the constant feedback to the authors made it a preferred outlet for contributions with high ambitions. It expanded from four to six issues a year in order to accommodate the flow of high-quality articles, but at the same time providing more space than many other journals.

This coincided with a period where journals also became the preferred form of publication. Early on in peace research history, working papers, in-house publications, and occasional works were seen to be enough. They constituted 'alternative' forms of publications. However, it was obvious that they were not read and had little impact. The same applied to dissertations, particularly in the Anglo-Saxon tradition where they were not even printed. However, evaluation of scholars was largely based on their publication record. Journal articles became an appealing option. The book publishers were also interested. Thus, academically oriented publishers went into journals. *JPR* and PRIO made an agreement in 1987 with Sage and the first issue with this publisher appeared in 1987. It was a path-breaking deal that was advantageous to all. Nils Petter was at the heart of this negotiation, resulting eventually in an income that made it possible to sustain editorial staff. Other journals were to follow, but *JPR* led the way.

The journal is also a leader in a number of other regards. Its readership has grown dramatically; today best indicated by downloads of journal articles around the world. The new measures that have been developed, notably impact factors, tell the same story, as do the number of submissions and the rejection rates. The importance for peace research to have a journal with that very name and with the recognition for quality cannot be exaggerated.

There are three examples of *JPR's* pioneering role: replication of data, publication of datasets, and data-based articles on novel topics. The first one concerned the demand that the authors provide replication data. This became a requirement for *JPR* authors in 1998 and today seems almost self-evident. If arguments and conclusions are based on data, and nobody is able to replicate the study, how can it by credible? Certainly there were embarrassing cases, published in other journals, when the author could not provide such information, as the original dataset had not been preserved. There seems still to be more space for actually doing such replication studies.

The second example is the data feature segment of the journal. One of the first such arrangements was the cooperation between the Uppsala Conflict Data Program (UCDP) and *JPR,* beginning with an article in 1993 on armed conflicts since 1989 (Wallensteen and Axell 1993). At this time, UCDP was strengthening its definitions, not the least by clarifying the notion of incompatibility as central in the concept of conflict. Thus, the agreement with *JPR* was a way to publish this new and more valid understanding. It led to further cooperation, when Nils Petter Gleditsch managed to secure a grant from the World Bank (negotiated with Paul Collier who was at the World Bank at the time), to 'backdate' armed conflicts from 1988 to 1946, thus creating a dataset for all conflicts since the end of the Second World War.[4] The work was done, and continues to be done, by UCDP (Themnér and Wallensteen 2014). It has retained the name, however, the UCDP/PRIO dataset, mostly in order to distinguish it from all the other UCDP data sets of conflicts, actors, peace agreement, etc. The annual publication of the updated version of this data article, after appropriate review, certainly has given *JPR* a wider readership and contributes to the attraction of the journal.

It has been followed by further publications of datasets, which, by today, is one of the marked features of *JPR*. Now, almost every issue introduces a new and significant collection of data. *JPR*, in other words, has been a leading actor in the 'data revolution' that today characterizes social sciences. I would even say *JPR* has pioneered this revolution as it pertains to issues of violence, war, and peace.

The third example, providing space for data-based articles on central issues of the time, is well reflected in this volume. Important recent findings in peace research have rested on data, notably the hypothesis of the democratic peace. It is undoubtedly one of the most debated notions in the past decades, and it seems the dust has still not settled. There are continuous challenges to the thesis that democratic states do not fight each other and have fewer civil wars. There are notions of a

---

[4]A point of departure was a list of candidate cases developed by Håvard Strand at PRIO, while the criteria and the final version was the work of the UCDP team, see Gleditsch, Wallensteen, Eriksson, Sollenberg and Strand (2002). For a history of UCDP see Wallensteen (2011: 111–124).

'capitalist' as well as a 'territorial' peace. *JPR*, and Nils Petter Gleditsch personally, have contributed to this discussion, as can be seen in Chaps. 4 and 8 in this volume.

A similarly data-based discussion is the one on whether wars are declining or not. Today, it is mainly associated with Steven Pinker (2011), who takes a long-term historical perspective, but also with the works of Andrew Mack (2005) and Joshua Goldstein (2011) who focus more narrowly on the post-World War II or post-Cold War periods. Again, this is a conversation where Nils Petter Gleditsch has participated (see Chap. 10) and it has not ended conclusively—and can it, ever?

In the 1990s, Nils Petter returned to the geographic variables that were important in his early days. At that time it was reflected in the study of air routes, but this time in terms of the connection of climate issues and conflict. A debate framed around 'environmental security' had emerged in the 1980s and PRIO was engaged in it early on, as was the Department in Uppsala. The 2000s saw even more attention to this issue, particularly with the publications from the *Inter-governmental Panel on Climate Change* (IPCC). *JPR*, PRIO and Nils Petter Gleditsch have all become leaders in this field of inquiry. These studies say that, as of now, there is no convincing evidence of a close connection between changes induced by climate variables and the onset of war, but there are many elements that need to be sorted out. This is definitely an issue that is of a lasting nature and will continue to be challenging. *JPR* is an important venue for this discussion. The articles on this topic are among the journal's most cited and Nils Petter's contribution in 2012 with a massive special issue of *JPR* with an introduction by himself (Chap. 9 in this volume) is at the very top of the league.

Certainly, this volume illustrates the multifaceted work of one of the pioneers of peace research. On many accounts, the fundamentals of peace research and of the scholarly work of Nils Petter Gleditsch have remained the same. There is an adherence to a precision in terminology, finding indicators that are not only reliable but also valid for the question at hand, searching for reproducible evidence that can be scrutinized by others, locating the relevant topics that will be important in the future for peace research. This is the Nils Petter that I met in the 1960s.

There are also some changes: Peace research now is firmly entrenched within universities, although the first chair in Scandinavia that was held by Galtung was lost to the field when he resigned in 1978. Nils Petter was appointed to Norwegian University of Science and Technology (NTNU) in Trondheim in 1993 and still retains an affiliation as professor emeritus. With the experiences from Uppsala in mind, I would say that creating chairs is the only way to establish peace research as a discipline within the universities and placing peace research on par with other forms of social science inquiry.

Leading an organization is another unexpected and notable achievement. Nils Petter Gleditsch held the position as President of the world's largest professional organization for international studies, International Studies Association in 2008–2009. Again, this was a remarkable achievement. Only three ISA Presidents had been based outside North America before this, and so far, none after his term. Nils Petter Gleditsch demonstrated that the North American domination could be broken. Furthermore, very few of the previous presidents have had such a close connection to organized peace research. Thus, Nils Petter's tenure was another founding experience.

These remarkable achievements have helped bring attention to the advances of peace research and its continued vitality. Debates and controversies remain, but today arguments are based on a stronger empirical footing. Nils Petter Gleditsch has been crucial in attaining this breakthrough. For this and all his other accomplishments those of us engaged in peace research are immensely grateful!

Uppsala, Sweden   Peter Wallensteen
January 2015

References

References for articles authored or co-authored by Nils Petter Gleditsch are found in the bibliography in Chap. 2.

Galtung, Johan, 1971: "A Structural Theory of Imperialism", in: *Journal of Peace Research*, 8,2: 81–117.

Goldstein, Joshua S., 2011: *Winning the War on War: The Decline of Armed Conflict Worldwide* (New York: Dutton).

Homans, George C., 1961: *Social Behavior: Its Elementary Forms* (New York: Harcourt, Brace, and World).

Mack, Andrew (Ed.), 2005: *Human Security Report 2005: War and Peace in the 21st Century* (New York: Oxford University Press).

Pinker, Steven, 2011: *The Better Angels of Our Nature* (New York: Viking).

Singer, J. David (Ed.), 1968: *Quantitative International Politics: Insights and Evidence* (New York: Free Press).

Themnér, Lotta; Wallensteen, Peter, 2014: "Armed Conflicts 1946–2013", *Journal of Peace Research*, 51,4: 541–554.

Wallensteen, Peter, 2011: *Peace Research: Theory and Practice* (London: Routledge).

Wallensteen, Peter; Axell, Karin, 1993: "Armed Conflicts after the Cold War", *Journal of Peace Research*, 30,3: 331–346.

Wiberg, Håkan, 1988: "The Peace Research Movement", in: Wallensteen, Peter (Ed.): *Peace Research: Achievements and Challenges* (Boulder: CO: Westview): 30–53.

Peter Wallensteen obtained his Ph.D. from Uppsala University in 1973 and served (1985–2012) as the first Dag Hammarskjöld Professor of Peace and Conflict Research in Uppsala, where he now is Senior Professor. Since 2006 he is also the Richard G. Starmann Sr. Research Professor of International Peace Studies, Kroc Institute, University of Notre Dame. He was the Head of the Department for Peace and Conflict Research, Uppsala University 1972–99. He is the Director of the Uppsala Conflict Data Program and leader of the Special Program on International Targeted Sanctions (SPITS), both at Uppsala University.

Issuing Officer's Signature, Rank and Appointment

For Supreme Commander, A.E.F.

Endorsements

VALID FOR OUTWARD JOURNEY ONLY.

[B44/618] 35000 10/44 W.O.P. 19371

On 16 May 1945, just eight days after the German occupation forces in Norway surrendered, the Military Permit Office in London issued a permit from the Allied Expeditionary Force for Nils Petter Gleditsch (3) 'to enter the Zone of the Allied Forces in NW Europe, Norway'. Apparently, he was exempted from signing.

Nils Petter Gleditsch (3) arrived in Norway for the first time on the British ship Andes on 30 May 1945, safe between his parents and accompanied by his aunt. *Photo* Unknown press photographer

Acknowledgments

Acknowledgments are given in the first footnote to each of the reprinted articles. My work in recent years has been supported by the Research Council of Norway through the core grant to PRIO and various project grants. I also acknowledge support for recent projects from KORO (Public Art Norway), the Norwegian Non-Fiction Literary Fund, and the Gløbius Fund. PRIO supported the purchase of open access for this volume. Three publishers—SAGE, Taylor & Francis, and Wiley—granted permission to reprint my journal articles. I am also grateful to Hans Günter Brauch for his entrepreneurship in academic publishing and for inviting this volume, to Peter Wallensteen for his overly complimentary preface, to Olav Bjerkholt, Ådne Cappelen, Mats Hammarström, and Henrik Urdal for comments on the introduction, to PRIO's librarian Odvar Leine for tracking down obscure articles and bibliographic details, to Maria Iselin Bjerkelund for competent technical assistance, and to Sarah Pettersen for copyediting. None of them bear any responsibility, etc. Halvard Buhaug set me straight on the noise level of the V2. And, of course, thanks to my family for (mostly) tolerating my excessive devotion to various projects, including this one.

With some advance planning, conferences can be combined with outdoor activities. A group of participants from a PRIO-sponsored conference in Nicosia went hiking in the Trodos mountains on 26 April 2006. *Left* to *right* (*front row*): Ragnhild Nordås, Andrew John Feltham, Halvard Buhaug, Helga Malmin Binningsbø, Kristian Skrede Gleditsch, David Cunningham, (*second row*): Jan Ketil Rød, Håvard Strand, Tove Grete Lie. *Photo* Nils Petter Gleditsch

Contents

Left-wing publishing still a boys' club. Editors at Pax publishing house, 1965. From *left* to *right*: Hans Fredrik Dahl, Tore Linné Eriksen, Nils Petter Gleditsch (holding a copy of his first book), Kjell G. Rosland, Theo Koritzinsky, and the publisher Tor Bjerkmann. *Photo* Aage Storløkken, Scanpix

The author stands trial for a national security violation, Oslo City Court, May 1981. *Photo* Arne Pedersen, Dagbladet

Part I
On Nils Petter Gleditsch

Nils Petter Gleditsch delivering his presidential address to the International Studies Association, San Francisco, USA, 27 March 2008. *Photo* Stein Tønnesson (PRIO)

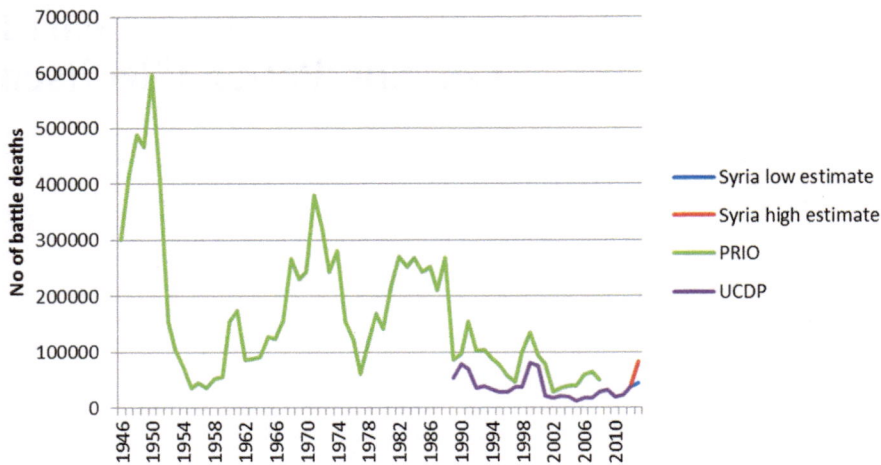

Graphs speak louder than words. A first version of the curve showing the decline in battle-related deaths (combatants as well as civilians) after World War II was first published in Lacina/Gleditsch (2005), based on the PRIO Battle Deaths Dataset generated by Bethany Lacina. It has been reproduced in numerous shapes and forms since then. This updated version shows the UCDP battle deaths data (1989–2013) and the PRIO data (1946–2008) in two separate time series. The Uppsala criteria for inclusion are slightly stricter, but the two series are roughly parallel for the 20 years that they overlap. No best estimate has yet been fixed for the casualties in Syria in 2013, so a high and a low estimate are given instead. This graph was created by Ida Rudolfsen for Gleditsch et al. (2015). The on-line appendix to that article includes a similar graph corrected for population size (i.e. depicting the probability of being killed in battle); it shows an even stronger decline of violence

Chapter 1
A Life in Peace Research

*Non, rien de rien. Non, je ne regrette rien.**

My personal life has been influenced by war in several ways. During the Spanish Civil War, my parents were active in the solidarity movement for the Republic and under the auspices of the Norwegian Committee for Spain they helped to support a hospital in Alcoy and a children's orphanage in Oliva. Although their efforts didn't succeed in saving the Republic, it had the inadvertent happy consequence of adding a little Spanish girl to our family—so when I was born, I had a lovely older sister. During the German invasion of Norway in April 1940, my parents were recruited to the transport of the gold reserves of the Norwegian Central Bank to safekeeping outside Norway, an operation led by my uncle. As a result, my parents ended up in England for the duration of the war. I was born in the London area in the summer of 1942 and spent some early years in the company of incoming V1s and V2s. Back in occupied Norway, two of my close relatives were tortured for their participation in resistance activity and a third was shot during the emergency in Trondheim when I was close to three months old (Berg 2007). Obviously, I don't remember any of this, so it would be an exaggeration to claim that experiencing war in my formative years influenced my choice of profession. My parents' political attitudes and activities were obviously more influential in leading me to join the peace movement and the labour movement.[1]

My professional life (I hesitate to think of it as a career) largely coincides with the history of peace research in Norway. I joined PRIO as a graduate student of

---

[1]My parents have described their experiences in Spain and the two-month campaign in Norway following the German invasion in Gleditsch/Gleditsch (1954). For the rescue of the Norwegian gold, (see Pearson 2015).

---

*Lyrics by Michel Vaucaire in 1956, as immortalized by Edith Piaf in 1960. Regrettably, her recording was dedicated to the French Foreign Legion http://en.wikipedia.org/wiki/Non,_je_ne_regrette_rien.

© The Author(s) 2015
N.P. Gleditsch, *Nils Petter Gleditsch: Pioneer in the Analysis of War and Peace*, SpringerBriefs on Pioneers in Science and Practice 29, DOI 10.1007/978-3-319-03820-9_1

sociology[2] in January 1964 and have remained there until today, with outside adjunct positions and shorter leaves of absence that have not really interrupted the continuity. Organized peace research started in Norway in the spring of 1959, with the formation of the Section for Conflict and Peace Research at the Institute for Social Research, which became PRIO in 1964 and a fully independent institute in 1966.[3]

1.1 The Beginnings

For the first five years I followed the activity at a discreet distance, fuelled mainly by my interest in the writings of Johan Galtung, PRIO's founder. I joined the Norwegian section of the War Resisters' International around 1959. I was greatly inspired by its pacifist manifesto (Galtung 1959) and became a conscientious objector. I am unable to put a date to my interest in social science. I grew up in a family with many natural scientists on my father's side, including an aunt who was a professor of chemistry and the second women professor in Norway.[4] Consistent with everyone's expectations, I enrolled as a science student at the University of Oslo in fall 1960, but didn't stay long. After (barely) finishing the introductory courses, I took a year off, including a four-month peace march from London to Moscow—the European part of the San Francisco to Moscow Walk for Peace.[5] My ambitious parents had put me in primary school a year early (as a result I was always the smallest kid in class until the final year of high school)[6]; now I 'wasted' that year. When I went back to the university, it was to study sociology, with minor degrees in philosophy and economics. Galtung (originally a statistician) and my father (a civil engineer and geographer) ganged up to try to persuade me to take at

---

[2]My 'defection' to political science (which I never studied) occurred much later.

[3]The Section was renamed Peace Research Institute, Oslo in 1964, although it remained part of ISR. In 1966, PRIO became fully independent and the name was changed to International Peace Research Institute, Oslo. The acronym was retained. This led to some confusion, which finally ended in 2010 when the name became Peace Research Institute Oslo (no comma!), with the banner 'Independent, International, Interdisciplinary' proudly displayed on the web pages. I will refer to the institute as PRIO throughout. For a history of PRIO, with an emphasis on its various engagements with Norwegian authorities, see Forr (2009).

[4]For brief biographies of Ellen Gleditsch, see www.physics.ucla.edu/ ~ cwp/Phase2/Gleditsch,_ Ellen@842345678.html and www.epigenesys.eu/en/science-and-you/women-in-science/773-ellen-gleditsch.

[5]See Lehmann (1966) and Lyttle (1966) for extensive descriptions from two of the participants and Wernicke/Wittner (1999) for a study by two peace movement historians.

[6]For a brief retrospective on my high school days, see Gleditsch (2003a).

least one minor in science, but I was more interested in Galtung's sociology and
peace research than in his intellectual starting-point. In retrospect, I realize that I
should have taken them more seriously and spent at least a full year studying
mathematics and statistics.

Box 1.1: Brief CV

Born 17 July 1942 in Sutton, Surrey, England by Norwegian parents.
Married (1966) to Kari Skrede, two children (1971, 1973), two grandchildren
(2008, 2010).

Educated at the University of Oslo and University of Michigan. Mag.art.
in sociology from University of Oslo (1968). Co-chair of the campaign
against the Norwegian doctoral degree (1970).

Worked at PRIO since 1964, as research professor since 1988, editor of
JPR (1975–76, 1983–2010). Professor of political science (part-time),
Norwegian University of Science and Technology (NTNU) (1993–2013).

Served on various editorial boards and committees; president of the
International Studies Association (2008–09).

Member of the Royal Norwegian Society of Sciences and Letters and the
Norwegian Academy of Science and Letters.

Awarded the Lewis F Richardson Lifetime Achievement Award, Award
for Outstanding Research (Møbius Prize) of the Research Council of Norway,
and the Lifetime Achievement Award of the Conflict Processes Section,
American Political Science Association.

Outside academia served as editor of *Pax* magazine (1962–65), president
of the Norwegian Students' Association (1966), chair of a local branch of the
Norwegian Labour Party (1980s and 1990s), chair of the board of Bestum
School Band (1986–88), and as a columnist in *Aftenposten* (1992–99) and in
Forskningspolitikk (2008).

For further details, see www.prio.org/staff/npg.

My first task as a research assistant at PRIO was in a project on nonviolent
national defense. Galtung (1965) contributed an important conceptual article and I
edited a volume in Norwegian featuring historical examples of nonviolent resistance (Gleditsch 1965a), but my contribution to the academic literature was modest
(Gleditsch 1968c). The impressive record of Norwegian civilian resistance to the
nazification of Norwegian society during World War II made the idea of building a
national post-occupation defense seem plausible. Less realistic (as I see it today)
was the idea of building a nonviolent national defense that could deter foreign
invasion. The project petered out after a year or so. But nonviolence has remained
close to my heart, and I am happy that the study on nonviolent resistance as an
alternative to armed insurgency has recently spawned a special issue of *Journal of
Peace Research* (Chenoweth/Cunningham 2013) and a new PRIO project that has
become part of the 'family business' (Rivera Celestino/Gleditsch 2013).

1.2 International Interaction

Although peace research has been most concerned[7] with negative interaction (i.e. armed conflict), the analysis of positive forms of international interaction occupied a central position in PRIO's research program in the first decade. My next project concerned international aviation. This became a source of some amusement to my friends, who argued that the Vietnam War was more important—a point that I didn't dispute. The study of international organization memberships (Skjelsbæk 1971), international diplomacy (Kvadsheim 1970), and related topics promised to inform us about the structure of the international system. During a stint of teaching sociology at FLACSO in Chile, Galtung had observed that in order to fly from one Latin American country to another, one frequently had to travel via the US. In his early work, hypotheses from one level of organization transferred seamlessly to another. The international airline network looked to Galtung like a 'feudal system'. The underlings related mainly to the overlord (known as the 'underdogs' and the 'top-dog' in his terminology, or the 'serfs' and the 'slave-master' among those a bit further left politically) and were prevented from organizing or even communicating between themselves. Data for my project were readily available in international airline guides. This became the topic of my master thesis (Gleditsch 1967, 1968d) and it also led me to some work in graph theory (Høivik/Gleditsch 1970), ably guided by my colleague Tord Høivik, who like Galtung had degrees in statistics as well as sociology. Chapter 3 on time differences and international interaction is one of my publications in this area.

A wider significance of international interaction patterns was that at the time 'positive peace' (as distinct from 'negative peace', the reduction of war) was somewhat vaguely defined as 'the integration of human society' (Galtung 1964: 2). Thus, studying international interaction patterns was a legitimate and important part of the study of peace. It was not until much later that peace research started taking an interest in the idea of liberal peace (Oneal et al. 1996) and connected positive and negative peace empirically. I don't think I saw this link clearly at the time. Neither do I remember being very concerned with 'the contact hypothesis', despite what Peter Wallensteen says in his preface. In fact, even after I became convinced that there was something to the idea of a democratic peace (Gleditsch 1992c) my argument was tainted with skepticism about the most pervasive form of positive peace: Could economic interdependence really have anything to do with reducing armed conflict?[8] My attitude was undoubtedly reinforced by the heated Norwegian debate on EU membership. I voted 'no' in both referenda (1972 and 1994) and in 1971–72 contributed to the fighting spirit of the anti-membership movement by a series of newspaper comments on the polls. In these articles, several colleagues and I argued (correctly, as it turned out) that there was a majority against Norwegian membership and that those who said otherwise were tweaking the data. We

---

[7]But not exclusively, cf Gleditsch et al. (2014).

[8]See Gleditsch (1995b) and Chap. 4, note 16.

eventually summed this work up in a book (Gleditsch/Hellevik 1977), but at the end of the day this probably had greater political than academic significance.

Returning for a moment to our early view of democracy, one of the early lessons from my mentor was to dismiss its importance for peace (Galtung 1967: 29). Galtung would later characterize democracies as particularly self-righteous and belligerent (Galtung 1996: Chap. 4), while I moved in the opposite direction. But prior to my conversion, when in an article (Gleditsch 1977) I listed five key 'global values', democracy (or even good governance) was not one of them.[9] We were more concerned with reversing dependency and promoting equality, a trait that was also reflected for many years in PRIO's governing structure. When Galtung stepped down as director in 1970, PRIO adopted a 'flat' governing structure, although it retained a position as director. The salary structure was amended in 1972 to a system where everyone was on the same ladder and you could only move up by seniority (Gleditsch 1974a, 1980). But contrary to our hope, the system was never copied by other scholarly institutions, not even by those that had adopted a more revolutionary rhetoric.[10] When a majority of the researchers tired of equality in practice, PRIO reverted to a traditional governing structure and salary system for the public sector.

1.3 The Peace Dividend and the Arms Race

Another aspect of my work concerns the peace dividend or, more broadly, the economic effects of disarmament. In the 1960s, developing countries pressed for diverting resources from the arms race to development aid and this concern was picked up by the UN. Early on, I played a minor role in helping to organize a major conference on this topic (Benoit 1967).[11] Much later, I joined forces with two economists from Statistics Norway and wrote several reports for the UN on disarmament and development, eventually resulting in two books on the peace dividend in Norway (Gleditsch et al. 1994) and world-wide (Gleditsch et al. 1996). The global project was conducted through Project Link, an international network for econometric modeling. It was a source of some pride that our 1996 book had a preface by one Nobel laureate (Wassily Leontief) and a chapter by another (Lawrence Klein). In retrospect, I am now inclined to believe that the key theme of linking development to disarmament was an intellectual and political mistake. But the many econometric studies conducted under this rubric certainly contributed to

---

[9]The five were welfare, peace, justice, pluralism, and ecological balance, largely copied from a colleague (Skjelsbæk 1973).

[10]In fact, many ultra-radical social scientists viewed our salary system as an implementation of 'cake theory'. This was left-wing jargon for a zero-sum theory of salaries and clear evidence of 'false consciousness'.

[11]In a combination of PRIO empire-building and generosity to a student, Galtung put my name on the title page in the same font as the editor's. But in fact I did little more than open the envelopes with the manuscripts of the various chapters and pass them on to the editor.

undermining facile notions that the arms race was necessary to avoid economic collapse in market economies. My work in this area was totally dependent on the insights of my economist colleagues. Although I wasn't completely a free rider, intellectually speaking, I chose not to reproduce any of this work here. A brief introduction can be found in Gleditsch et al. (1983) and my most recent effort (in Norwegian) is Cappelen/Gleditsch (2012).

The economics of disarmament is, of course, closely related to the arms race. The Richardson (1960) model of arms races is a true classic in peace research and is mostly tested by using arms expenditures as the indicator of arming. This model was challenged in peace research in the 1970s, notably by Senghaas (1979) who thought of the arms race as inner-directed or 'autistic'. A PRIO-sponsored conference assessed this debate, leading to an edited volume (Gleditsch/Njølstad 1990). My own modest contribution was an introductory article (Gleditsch 1990b). In a review of the book, Andrew Mack generously suggested that my introduction was something that 'students would kill for',[12] but no casualties have been reported.

1.4 Norway in the Nuclear Arms Race

I have spent a lot of time (probably too much) looking into the Norwegian contribution to the military side of the East-West conflict. In a sense, this followed naturally from my early work in the peace movement.[13] Norway's controversial decision to break with neutralism and join NATO in 1949 was accompanied by a declaration that foreign bases would not be established on Norwegian soil. This was meant to assuage Soviet fears that Norwegian territory might become a springboard for offensive attacks. The establishment of joint NATO commands after the outbreak of the Korean War put this policy under some pressure. But in 1961, after another heated debate, the social democratic government decided that Norway would remain nuclear-free. However, through NATO's infrastructure program and bilateral arrangements with the US, Norway was littered with foreign-funded military electronic installations (radars, transmitters for navigation signals and communication, and stations for intelligence gathering), as well as airfields, all of which seemed likely to be employed in a nuclear showdown (Gleditsch/Lodgaard 1977). Some parts of this infrastructure had a clear role in the defense of Norwegian territory. Others did not, and could in my view only be understood as an exchange of strategic services: Norway gained in security by sheltering under the nuclear umbrella of the US, and in turn agreed to contribute to its strategic war-fighting capability. Some installations that served as peacetime preparations for strategic

---

[12]While Andy recalls having written something to this effect (personal communication 28 November 2014), neither of us has been able to locate the source.

[13]Notably in the peace movement magazine *Pax* and later the publishing house with the same name, whose history has just been written (Helsvig 2013).

nuclear warfare were likely to be taken out in wartime by the Soviet Union, thus bringing the war to Norway even in the absence of a direct challenge to Norwegian sovereignty. Such an exchange of strategic services remained controversial domestically, particularly in the Labour Party. The Norwegian government therefore maintained a high level of secrecy and blurred the line between joint NATO projects and bilateral US operations. Of my work in this area, I am most proud of collaborative efforts with Owen Wilkes, an unorthodox Kiwi with few formal credentials but wide knowledge of the relevant science and an extraordinary knack for extracting information from the publications of the national security establishments. In a book on military navigation (Wilkes/Gleditsch 1987), we showed that the two transmitters for the Loran-C navigation system were built in Norway in 1960 at the request of the US and in preparation for the deployment of Polaris ballistic missile submarines in the Norwegian Sea. It had taken 15 years to reveal this to the Norwegian public, and even then only through investigative research and the leaking of a secret public inquiry (Schei et al. 1977). The military and political establishments were not pleased with our work. They would have been even less so, had they realized that it was I who obtained the secret report from a left-wing parliamentarian and passed it to the publisher. Indeed, the possibility of prosecuting us was considered, but no legal basis was found.

Eighteen months later, however, when we published a report on US-funded intelligence stations in Norway (Wilkes/Gleditsch 1979, 1981), the establishment struck. We were charged with a national security violation and eventually given a 6 months jail term, suspended by a 3:2 vote on the Norwegian Supreme Court (Gleditsch 1981a–c, 1982). The legal basis for the conviction was the so-called 'puzzle doctrine', originally formulated in a 1954 spy case. Although we claimed to have worked only with open sources, the combination of such sources could be detrimental to national security, according to the Court. At the time, the Conservative Party had just won the election and had committed itself to a public inquiry of what was going on at PRIO. But the committee appointed to look into the issue, tackled it in an academic manner (Midgaard et al. 1985), and the consequences for PRIO were limited. However, as a convicted felon I was denied entry into the US for several years—a somewhat ironic outcome given the stronger freedom-of-information tradition in the US. Unusually for a Norwegian, I still need a visa to enter the US, and mine still has several footnotes that have led me into interesting conversations with immigration officials at various airports.

1.5 Secrecy and Espionage

My encounter with official secrecy in Norway spurred an interest in secrecy legislation in other countries. I obtained a grant for a comparative study of freedom-of-information legislation in Norway and the US. A young lawyer, later to become head of Norway's Economic Crimes Unit, wrote a book on this topic (Høgetveit 1981) and, with me, an article in English (Gleditsch/Høgetveit 1984).

When the Arne Treholt spy case broke in January 1984, I was naturally interested. He and I were of the same generation on the left side of student politics and his arrest as a spy for the Soviet Union came as an unpleasant surprise to all who knew him. In his initial police interrogation he confessed to being caught in a trap well known from the world of espionage: you let an intelligence agent get too close, and eventually it becomes difficult to pull out. Most of us had been exposed to Soviet agents, who were very active among young people on the left in the 1960s, but we had kept our distance or got out in time. Treholt did not. He went on to have a political career that must have exceeded his handlers' expectations by a wide margin. Soon after his arrest, he withdrew his first statements and switched to an argument that he had only conducted private diplomacy—well hidden to the Foreign Ministry, his friends, and his family. The limited evidence for his leaking of material that was seriously detrimental to national security led to further controversy and spawned a host of conspiracy theories. I classified the relevant writings into traditionalist, revisionist, and post-revisionist, according to a scheme often used by Cold War historians (Gleditsch 1994b, 1995e). The shorter English article is Chap. 5 in this volume. The case remains controversial even today. A large number of new books have appeared, but most of them have contributed more myths than substance.

Another reason for my concern about the Treholt case was that some of his defenders drew a parallel between his 'unofficial diplomacy' and work undertaken in research institutes. Given my own conviction under some of the same paragraphs in the penal code, I was anxious not to let this view stand unopposed. In a short newspaper article (Gleditsch 1985), I warned against equating research conducted in full openness with covert contact with intelligence officials. I argued that 'if for years you behave like a spy, you must reconcile yourself to the idea that someone might think that in fact you are one.' Much to my surprise, this has over the years become my most frequently cited 'bon mot' in Norwegian media. It must have touched a raw nerve since Treholt (2004: 324, 476) cites it twice (albeit inaccurately).

Fear of foreign espionage and distrust of left-wing opposition were prime and intertwined motives behind political surveillance during the Cold War. When a window of opportunity appeared in 1999, I applied for my 'secret file' and eventually obtained a substantial sum in reparations for illegal collection and recording of information about my activities (Gleditsch 2003c). The legal window closed in 2007, and currently no private individuals can apply for their Norwegian intelligence records, however outdated (Gleditsch 2014b). Although measures for the monitoring of the secret services have been strengthened, traditional habits of official secrecy are hard to change, and today's information society invites private as well as official invasions of privacy. An application for my Stasi file is still pending and I hope at some point to apply for my FBI file. But perhaps I should be more concerned about my file at Google.

1.6 The Waning of War

While peace research initially had an optimistic view of the potentialities of social science, it was quite pessimistic about the state of the world. In part, peace research was founded in protest against the nuclear arms race and the dominant position of realist thinking in international relations. Galtung (1964, 1967) had written that the twentieth century was the bloodiest in human history and that the world was moving in the direction of more conflict. There was widespread fear that human civilization might not survive to see the new millennium. The number of armed conflicts was rising through most of the Cold War. Rummel (1994) had not yet taught us that the victims of 'democide' (violence by governments against unorganized civilians, mostly their own citizens) by far outnumbered the victims of war. Had we been more alert and attuned to this form of violence, we would have worried that the world had just experienced its largest outburst of what we now call one-sided violence. This was 'the Great Leap Forward' in China, with 45 million deaths over just 4 years (1958–61) (Dikötter 2010). We now know that the rising number of conflicts in part was an artifact of the decolonization process and the creation of new and fragile states, that the periodic peaks in the lethality of war generated by the Korean War, the Vietnam War etc. were progressively lower and never reached the levels of violence of the World Wars, and that despite Cambodia (1975), Rwanda (1994), and Darfur (2003), one-sided violence is also on the wane.[14] It was not until the end of the Cold War that we heard Mueller (1989) arguing that war was obsolescent, and eventually our own data collection efforts (Gleditsch et al. 2002; Lacina/Gleditsch 2005) helped to persuade us that this was so. In fact, as Payne (2004) and Pinker (2011) have argued, violence is increasingly getting discredited as a tool in human interaction. Looking back at the bloody history of the human race, Pinker and others have even cast serious doubt on the top rank of the twentieth century on violence if we adjust for population size, in other words, look at the probability that a random individual will get killed in violence. Curiously, many peace activists remain skeptical of the good news. My own view is that the peace movement (in which I am no longer very active) should claim its part of the credit for the things we have done right. Chapter 10 contains my introduction to a symposium on Pinker's book, based on a session at the International Studies Association convention in 2012.

Another part of my generation's introduction to peace research was that wars are increasingly harmful to participants and to civilians. A popular saying in peace research circles ran roughly like this: In the old days, 90 % of those killed in war were combatants, today 90 % of those killed are civilians.[15] As Roberts (2010) and Eck (2005) have shown, there is absolutely no scientific basis for these figures,

---

[14]See Eck/Hultmann (2007) and annually updated data from the Uppsala Conflict Data Program at www.pcr.uu.se/research/ucdp/datasets/.

[15]Or, according to Nobel physicals laureate Max Born (1964: 4), from 95/5 in World War I over 50/50 in World War II to 16/84 in the Korean War—no sources provided.

though they are still repeated from time to time. We will probably never have accurate statistics on this, but a more reasonable hypothesis seems to be that the ratio of civilian to military casualties remains roughly stable over time, although it probably varies from one conflict to another.

My embracing the waning-of-war argument has led to occasional accusations of excessive optimism. As noted, peace research was born in an era of considerable pessimism—reasonable enough when living under a 'balance of terror'. But after rereading the articles I have selected for this volume, I readily plead guilty to the charge of optimism. While the technological enthusiasm of Chap. 3 may now seem excessive, I have only been strengthened in the environmental optimism that shines through in Chaps. 6, 9, and 11, as well as in my more recent belief in the global decline of violence as set out in Chaps. 8 and 10. The two are related, since environmental degradation (and climate change in particular) is a main contender to fill the gap in what Mueller (1994) aptly calls the 'catastrophe quota'.

Of course, we should not be too discouraged when we are proven wrong about things we believed in during the pioneer period. Indeed, what would be the point of doing research if we could only confirm what we already knew? Too many social scientists are satisfied by uttering periodic post hoc statements to the effect that 'I am not surprised'. This does not advance the research frontier. We need scholars who stick their neck out with bold hypotheses and even venture into predictions (Hegre et al. 2013), even if at first they are not likely to be spot-on or even near misses.

I am happy to report that there was one widespread myth that we did not embrace. According to an obscure study, there had been only 292 years of peace since 3600 BC and 14,531 wars had claimed 3,640,000,000 lives. Moreover, 1,656 arms races had been conducted and all but 16 of them had ended in war. In the late 1950s and early 1960s these numbers made their way through countless newspapers, military journals, and peace movement periodicals. We were warned off this story by a memorandum from the Rand Corporation (Haydon 1962) that revealed that the these 'calculations' were cited from an article by Norman Cousins (1953) that reported on 'an imaginary experiment' conducted by 'Dr. P. Storhjerne' (Dr. P. Great Brain), a former President of the Norwegian Academy of Science. J. David Singer, founder of the Correlates of War project, tipped us about this memo when he spent a sabbatical in Oslo in 1963–64.[16] Later, we advised *Readers' Digest* not to give further publicity to this tale when (to their credit) they looked for Norwegian sources to verify it.

---

[16]See Singer/Small (1972: 10–11) and Gleditsch (1964). Jongman/van der Dennen (1988) were able to trace the figures used by Cousins back all the way to 1864. Despite the debunking, Fine/O'Neill (2010) report the US Secretary of the Navy using the figures as late as 2007.

1.7 Publish or Perish

A high point in the pioneer period was the launch of *JPR* in 1964. The journal was originally planned as *Journal of Peace and Conflict Research,* in line with PRIO's own name and as a concession to those who still remained skeptical of 'peace' as an academic topic. However, Galtung finally settled for *Journal of Peace Research* and deleted 'conflict' from PRIO's name. The first publisher, Norwegian University Press, was probably more skeptical about the economic prospects than about the name, and PRIO remained responsible for the finances of the journal. For over 22 years we (or rather the Research Council of Norway) paid them a subsidy for their publication services. Not until we put the journal out for competitive bidding on the international market in 1986, did we realize that, at least by then, the publisher should be paying us! Today, owning *JPR*, or any reputable journal, is a profitable enterprise.

Shortly after having been present at the creation, I was enlisted as an assistant editor in 1965 and served as editor for some 28 years (1975–76, 1983–2010). I also had the privilege of handpicking my successor, Henrik Urdal. Under his steady hand the journal has continued to go from strength to strength (Urdal et al. 2014). Over the years, I have written various shorter pieces about the use of referees (Gleditsch 1989c), the most-cited articles in *JPR* (Gleditsch 1993), the pros and cons of double-blind reviewing (Gleditsch 2002b), the importance of posting replication data (Gleditsch et al. 2003), open access Gleditsch (2012f), and the gender gap (Østby et al. 2013). Our article on replication showed empirically that those who posted their data were more likely to become famous (or at least be cited more frequently) than to be scooped. *JPR* also acquired a reputation as a leader in the replication movement. The gender article revealed a continuing but lessening gender gap among *JPR's* authors, but no evidence of gender discrimination. Obviously one cannot serve as journal editor for such a long time without spending a lot of time pondering the importance of publishing and Chap. 7 on *JPR's* review process is one example.

One of the things I learned from my mentor was the importance of publishing at an early age, and in visible channels. My first *JPR* article (Gleditsch 1967) was published when I was 25[17]—although in retrospect I can see that it wouldn't have hurt to have had it subjected to outside review (not common at the time) and leave at least a few more weeks for revision and polishing. Much later, I was able to publicize Erich Weede's publication law ('my personal rejection rate in journals is about 50 % and if I fall below that level, I haven't been sufficiently ambitious in where I submit my articles'), after its author had forgotten it (Schneider 2005: 258). Although any incentive system has its share of perverse outcomes, I am basically favorable to the Norwegian system for funding based on publication units (Gleditsch 2007a, b).

---

[17]I was beaten by Peter Wallensteen and Raimo Väyrynen, both 23 when publishing in *JPR* for the first time.

1.8 About This Volume

My nine articles reprinted in this volume were written over a period of 40 years (1974–2014). I decided to reprint only single-authored work. Obviously, much of my most important and certainly my most highly-cited work was published in jointly authored articles, such as Gleditsch et al. (2002) and Hegre et al. (2001). I strongly support collaborative writing and joint publication and have co-authored with over 100 different people. However, I don't want to appear here as a peacock in borrowed garb, so I accept the risk of being exposed as a mere jackdaw (Aesop, appr 600 BC/1820: 53). In this introduction, I have tried to place these articles in a broader context. They are republished here in the original form, except that I have corrected obvious typos and spelling errors and amended the citations and references to current *JPR* style. Where appropriate, I have also added references to a few published articles that were originally cited as unpublished papers. A comparison of the years of publication of the cited and the citing article will make it obvious where references have been added or updated. Beyond that, my work appears as it was originally published, warts and all.

1.9 Paying My Debts

I am grateful to Johan Galtung, founder of PRIO and *JPR,* for bringing me and many other young people into peace research 50 years ago. We were offered access to a demanding and stimulating research environment. It will be obvious to the discerning reader that Galtung and I have gone our separate ways on a number of issues. I tend to think that I am still faithful to important values that Galtung taught me in the mid-1960s, while he probably leans to the view so beautifully expressed in a the title of a Norwegian science fiction novel (Bringsværd 1974), *The one who has both feet planted on the ground stands still.* I am also grateful to my young Norwegian and Nordic colleagues in the pioneer days, several of whom became life-long associates and friends. Because they were so many, I hesitate to name any, but I make exceptions for three life-long collaborators, Peter Wallensteen (author of the preface to this volume), Andrew Mack (an Australian, but also an honorary Nordic), and Håkan Wiberg (former director of the Copenhagen Peace Research Institute, who sadly passed away in 2010).

Since the revival of a more traditional governing style at PRIO in 1986, four successive directors (Sverre Lodgaard, Dan Smith, Stein Tønnesson, and currently Kristian Berg Harpviken) have maintained the tradition of encouraging researchers at PRIO to develop their own projects and allowing a wide freedom of expression, with obvious (but perhaps not always visible or acknowledged) benefits for my own work.

For many years (and particularly since joining the faculty at NTNU in Trondheim on a part-time basis in 1993) I have been privileged to work with a large number of gifted students and younger colleagues. With their peers the world over, many of them are now taking the analysis of war and peace to new heights. This book is collectively dedicated to them.[18] I don't want to claim a major share of the credit for their achievements, but I hope I have paid back to the collective some of what I owe to my own mentors.

Before my time, several close relatives succumbed to tuberculosis. Another uncle, a medical doctor, became a leading figure in the Norwegian fight against the disease. Towards the end of his life, he witnessed the closing of one after another of his old hospitals and the virtual eradication of TB in Norway. Even if it would be premature to claim an equally happy ending to my own professional life as a student of conflict and peace, I'm hopeful that the next generation of scholars can record another notch in the waning of war.

References

To save space, I have omitted references to my own publications, which can be found in the bibliography in Chapter 2.

Aesop (attributed) (appr. 600 BC) The Bird in Borrowed Feathers. Reprinted in Bewick, Thomas; Bewick, John, 1820: *Select Fables: With Cuts* (Newcastle: S. Hodgson): 53–54. https://archive.org/stream/selectfableswith00bewi#page/52/mode/2up.

Benoit, Emile (Ed.), 1967: *Disarmament and World Economic Interdependence* (Oslo: Norwegian University Press).

Berg, Thoralf, 2007: *Henry Gleditsch – skuespiller, teatergründer, motstandsmann* [Henry Gleditsch: Actor, Theater Gründer, Resistance Man] (Trondheim: Communicatio).

Born, Max, 1964: "What is Left to Hope For?", in: *Bulletin of the Atomic Scientists*, 20,4: 2–5.

Bringsværd, Tor Åge, 1974: *Den som har begge beina på jorda står stille* [The One Who Has Both Feet Planted on the Ground Stands Still] (Oslo: Gyldendal).

Buhaug, Halvard; Gates, Scott; Hegre, Håvard; Strand, Håvard; Urdal, Henrik, 2009: "Nils Petter Gleditsch: A Lifetime Achiever", in: *European Political Science*, 8,1: 79–89.

Chenoweth, Erica; Cunningham, Kathleen Gallagher, 2013: "Understanding Nonviolent Resistance: An Introduction", in: *Journal of Peace Research*, 50,3: 271–276.

Cousins, Norman, 1953: "Electronic Brain on War and Peace: A Report of an Imaginary Experiment", in: *St Louis Post-Dispatch*, 13 December: 13.

Dikötter, Frank, 2010: *Mao's Great Famine. The History of China's Most Devastating Catastrophe, 1958–62* (London: Bloomsbury).

Eck, Kristine, 2005: "Getting It Wrong: The 'Urban Myth' About Civilian War Deaths", in: Mack, Andrew (Ed.): *War and Peace in the 21st Century. Human Security Report* (Oxford: Oxford University Press): 75.

---

[18]Some of them have even been kind enough to publish a quantitative analysis of my work (Buhaug et al. 2009) and thereby, probably without any deliberate intention to manipulate the data, saving some of my early articles from appearing with zero citations.

Eck, Kristine; Hultmann, Lisa, 2007: "One-Sided Violence Against Civilians in War: Insights from New Fatality Data", in: *Journal of Peace Research*, 44,2: 233–246.

Fine, Gary Allan; O'Neill, Barry, 2010: "Policy Legends and Folklists: Traditional Beliefs in the Public Sphere", in: *Journal of American Folklore*, 123,488: 150–178.

Forr, Gudleiv, 2009: *Strid og fred: fredsforskning i 50 år: PRIO 1959–2009* [Strife and Peace. Peace Research for 50 Years. PRIO 1959–2009] (Oslo: Pax).

Galtung, Johan, 1959: *Forsvar uten militærvesen* [Defense without a Military] (Oslo: Folkereisning mot krig).

Galtung, Johan, 1964: "An Editorial", in: *Journal of Peace Research*, 1,1: 1–4.

Galtung, Johan, 1965: "On the Meaning of Nonviolence", in: *Journal of Peace Research*, 2,3: 228–257.

Galtung, Johan, 1967: *Fredsforskning* [Peace Research] (Oslo: Pax).

Galtung, Johan, 1996: *Peace by Peaceful Means: Peace and Conflict, Development and Civilization* (London: Sage).

Gleditsch, Nini; Gleditsch, Kristian, 1954: *Glimt fra kampårene* [Glimpses from the Years of Struggle] (Oslo: Dreyer).

Haydon, Brownlee W., 1962: *The Great Statistics of War Hoax* (Report. Santa Monica, CA: RAND).

Hegre, Håvard; Nygård, Håvard Mokleiv; Strand, Håvard; Urdal, Henrik; Karlsen, Joakim, 2013: "Predicting Armed Conflict, 2010–2050", in: *International Studies Quarterly*, 57,2: 250–270.

Helsvig, Kim G., 2013: *PAX forlag 1964–2014. En bedrift* [PAX Publishing House 1964–2014. An Enterprise] (Oslo: Pax).

Høgetveit, Einar, 1981: *Hvor hemmelig? offentlighetsprinsippet i Norge og USA, særlig med henblikk på militærpolitiske spørsmål* [How Secret? The Freedom of Information Principle in Norway and the US, with Particular Regard to Questions of Military Policy] (Oslo: Pax).

Jongman, Albert J.; van der Dennen, Johan M.G., 1988: "The Great 'War Figures' Hoax: An Investigation in Polemomythology", in: *Bulletin of Peace Proposals*, 19,2: 197–203.

Kvadsheim, Reidar, 1970: *The Diplomatic Network.* MA Thesis, University of Oslo (Oslo: PRIO).

Lehmann, Gerald, 1966: *We Walked to Moscow* (Raymond, NH: Greenleaf).

Lyttle, Bradford, 1966: *You Come with Naked Hands: The Story of the San Francisco to Moscow March for Peace* (Raymond, NH: Greenleaf).

Midgaard, Knut; et al., 1985: *Forskning om sikkerhets- og fredsspørsmål og internasjonale forhold* [Research on Issues of Peace and Security and International Relations]. NOU 1985: 17 (Oslo: Norwegian University Press).

Mueller, John, 1989: *Retreat from Doomsday* (New York: Basic Books).

Mueller, John, 1994: "The Catastrophe Quota: Trouble after the Cold War", in: *Journal of Conflict Resolution*, 38,3: 355–375.

Oneal, John R.; Oneal, Frances H.; Maoz, Zeev; Russett, Bruce, 1996: "The Liberal Peace: Interdependence, Democracy, and International Conflict, 1950–85", in: *Journal of Peace Research*, 33,1: 11–28.

Payne, James L., 2004: *A History of Force* (Sandpoint, ID: Lytton).

Pearson, Robert, 2015: *Gold Run: The Rescue of Norway's Gold Bullion from the Nazis, 1940* (Havertown, PA: Casemate).

Pinker, Steven, 2011: *The Better Angels of Our Nature* (New York: Viking).

Richardson, Lewis F., 1960: *Arms and Insecurity: A Mathematical Study of the Causes and Origins of War* (Pittsburgh, PA: Boxwood/Chicago, IL: Quadrangle).

Rivera Celestino, Mauricio; Gleditsch, Kristian Skrede, 2013: "Fresh Carnations or All Thorn, No Rose? Nonviolent Campaigns and Transitions in Autocracies", in: *Journal of Peace Research*, 50,3: 385–400.

Roberts, Adam, 2010: "Lives and Statistics: Are 90 % of War Victims Civilians?", in: *Survival*, 52,3: 115–136.

Rummel, Rudolph J., 1994: *Death by Government* (New Brunswick, NJ: Transaction).

Schei, Andreas; et al., 1977: *Loran C og Omega. Innstilling fra utvalget til undersøkelse av saken om etablering av Loran C og Omega-stasjoner i Norge* [Loran-C and Omega. Report from the

Committee to Investigate the Establishment of Loran-C and Omega Stations in Norway] (Oslo: Pax) [Originally issued as a classified report in 1975].

Schneider, Gerald, 2005: "Erich Weede: A Nonconformist Conflict Researcher", in: *European Political Science*, 4,3: 256–262.

Senghaas, Dieter, 1979: "Arms Dynamics and Arms Control in Europe", in: *Bulletin of Peace Proposals*, 10,1: 8–19.

Singer, J. David; Small, Melvin, 1972: *The Wages of War 1816–1965: A Statistical Handbook* (New York: Wiley).

Skjelsbæk, Kjell, 1971: "Growth of International Nongovernmental Organization in the Twentieth Century", in: *International Organization*, 25,3: 420–442.

Skjelsbæk, Kjell, 1973: "Value Incompatibilities in the Global System", in: *Journal of Peace Research*, 10,4: 341–354.

Treholt, Arne, 2004: *Gråsoner* [Grey Areas] (Oslo: Gyldendal).

Wernicke, Günter; Wittner, Lawrence S., 1999: "Lifting the Iron Curtain: The Peace March to Moscow of 1960–1961", in: *International History Review*, 21,4: 900–917.

Wiberg, Håkan, 1988: "The Peace Research Movement", in: Wallensteen, Peter (Ed.): *Peace Research: Achievements and Challenges* (Boulder, CO: Westview), 30–53.

Chapter 2
Bibliography

The bibliography is ordered chronologically within four general categories. I have included more or less all my academic work, broadly defined. Thus, the list includes some popular and even polemical articles and books that deal with topics on which I have also done research. I have also included a few writings that deal with other topics if they are cited in my personal retrospective in Chap. 1. I have omitted all other such writings, such as the history of the local school band (where my children once played) and my introduction to a Norwegian volume of Donald Duck comics. To save space, I have replaced my own name with my initials, NPG.

2.1 Books and Guest-Edited Special Journal Issues and Symposia in English

NPG; Leine, Odvar; Holm, Hans-Henrik; Høivik, Tord; Klausen, Arne Martin; Rudeng, Erik; Wiberg, Håkan (Eds.), 1980: *Johan Galtung. A Bibliography of His Scholarly and Popular Writings 1951–80* (Oslo: PRIO).

Wilkes, Owen; NPG, 1987: *Loran-C and Omega. A Study of the Military Importance of Radio Navigation Aids* (Oslo & Oxford: Norwegian University Press & Oxford University Press).

NPG; Njølstad, Olav (Eds.), 1990: *Arms Races—Technological and Political Dynamics* (London: SAGE).

NPG (guest Ed.), 1992a: "Defence Spending after the Cold War", in: *Cooperation and Conflict.* Special Issue 27,4: 323–441.

NPG; Bjerkholt, Olav; Cappelen, Ådne, 1994: *The Wages of Peace. Disarmament in a Small Industrialized Economy* (London: SAGE).

NPG; Risse-Kappen, Thomas (Eds.), 1995: "Democracy and Peace", in: *European Journal of International Relations.* Special Issue 1,4: 405–574.

NPG; Bjerkholt, Olav; Cappelen, Ådne; Smith, Ron P.; Dunne, J. Paul (Eds.), 1996: *The Peace Dividend.* In: Contributions to Economic Analysis (Amsterdam: North-Holland).

© The Author(s) 2015
N.P. Gleditsch, *Nils Petter Gleditsch: Pioneer in the Analysis of War and Peace*, SpringerBriefs on Pioneers in Science and Practice 29, DOI 10.1007/978-3-319-03820-9_2

NPG; Brock, Lothar; Homer-Dixon, Thomas; Perelet, Renat; Vlachos, Evan (Eds.), 1997: *Conflict and the Environment*. NATO ASI Series 2, Environment 33 (Dordrecht: Kluwer Academic).

NPG; Lindgren, Göran; Mouhleb, Naima; Smit, Sjoerd; de Soysa, Indra (Eds.), 2000: *Making Peace Pay: A Bibliography on Disarmament and Conversion* (Claremont, CA: Regina).

Diehl, Paul; NPG (Eds.), 2001: *Environmental Conflict* (Boulder, CO: Westview).

Schneider, Gerald; Barbieri, Katherine; NPG (Eds.), 2003: *Globalization and Armed Conflict* (Lanham, MD: Rowman & Littlefield).

Nordås, Ragnhild; NPG (guest Eds.), 2007: "Climate Change and Conflict", in: *Political Geography*, 26,6: 627–735.

NPG; Schneider, Gerald; Carey, Sabine (guest Eds.), 2010: "Exploring the Past, Anticipating the Future. Presidential Symposium", in: *International Studies Review*, 12,1: 1–104.

NPG; Schneider, Gerald (guest Eds.), 2010: "A Capitalist Peace?", in: *International Interactions*, 36,2: 107–213. Revised and expanded version as Schneider, Gerald; NPG (Eds.), 2013: *Assessing the Capitalist Peace* (London: Routledge).

Schneider, Gerald; NPG; Carey, Sabine (guest Eds.), 2011: "Prediction and Forecasting", in: *Conflict Management and Peace Science*, 28,1: 5–85.

Bernauer, Thomas; NPG (guest Eds.), 2012: "Events Data in the Study of Conflict", in: *International Interactions*, 38,4: 375–569.

NPG (guest Ed.), 2012a: "Climate Change and Conflict", in: *Journal of Peace Research*, 49,1: 1–267.

NPG (guest Ed.), 2012b: "Open Access in International Relations", in: *International Studies Perspectives*, 13,3: 211–234.

NPG; Pinker, Steven; Thayer, Bradley A.; Levy, Jack S.; Thompson, William R., 2013: "The Forum: The Decline of War", in: *International Studies Review*, 15,3: 396–419.

2.2 Books in Nordic Languages

NPG (Ed.), 1965a: *Kamp uten våpen* [Struggle without Arms] (Oslo: Pax). Revised Swedish edition, 1971: *Kamp utan vapen* (Stockholm: Prisma).

NPG, 1970a: *Norge i verdenssamfunnet: En statistisk håndbok* [Norway in the World Community: A Statistical Handbook] (Oslo: Pax). Revised edition, 1988.

NPG; Lodgaard, Sverre, 1970: *Krigsstaten Norge* [Norway—A Warfare State] (Oslo: Pax).

NPG; Østerud, Øyvind; Elster, Jon (Eds.), 1974: *De utro tjenere. Embetsverket i EF-kampen* [Unfaithful Servants. The Civil Service in the Common Market Struggle] (Oslo: Pax).

NPG; Hellevik, Ottar, 1977: *Kampen om EF* [The Common Market Struggle] (Oslo: Pax).

NPG; Lodgaard, Sverre; Wilkes, Owen; Botnen, Ingvar, 1978: *Norge i atom-strategien* [Norway in the Nuclear Strategy] (Oslo: Pax).

Wilkes, Owen; NPG, 1981: *Onkel Sams kaniner. Teknisk etterretning i Norge* [Uncle Sam's Rabbits. Technical Intelligence in Norway] (Oslo: Pax).

Botnen, Ingvar; NPG; Høivik, Tord, 1983: *Fakta om krig og fred* [Facts about War and Peace] (Oslo: Pax). Also as volume 7 of *Pax Leksikon*.

NPG; Møller, Bjørn; Wiberg, Håkan; Wæver, Ole, 1990: *Svaner på vildveje? Nordens sikkerhed mellem supermagtsflåder og europæisk opbrud* [Lost Swans? Nordic Security between Superpower Fleets and a European Departure] (Copenhagen: Vindrose).

NPG; Enckell, Pehr; Burchard, Jørgen (Eds.), 1994: *Det vitenskapelige tidsskrift* [The Scientific Journal] (Copenhagen: Nordic Council of Ministers).

NPG (Ed.), 1998: *Det nye sikkerhetsbildet. Mot en demokratisk og fredelig verden* [The New Security Picture. Toward a Democratic and Peaceful World] (Trondheim: Tapir).

2.3 Articles

NPG, 1964: "Krigens statistikk [The Statistics of War]", in: *Pax*, 3,9: 174.

NPG, 1965b: "Ikkevoldsforsvar som strategisk avverge [Nonviolent Defense as a Strategic Deterrent]", in: *Pax*, 4,1: 16–19, 32. Reprinted in: Koritzinsky, Theo (Ed.), 1967: *Alternativer – fem års fredsdebatt* (Oslo: Pax): 123–133; and in: Bakke, Tormod; Nilsen, Tom (Eds.), 1987: *Ikkevold – teori og praksis* (Oslo: Folkereisning mot krig): 136–143.

NPG, 1967: "Trends in World Airline Patterns", in: *Journal of Peace Research*, 4,4: 366–408. Slightly revised version with Galtung, Johan, 1980: "International Air Communication: A Study in World Structure", in: Galtung, Johan (Ed.): *Essays in Peace Research, Peace and World Structure* (Copenhagen: Ejlers): 152–204.

NPG, 1968a: "Hvem blir studentopprørere – og hvorfor? [Who Become Student Rebels—and Why?]", in: *Pax*, 7,6/7: 179–185.

NPG, 1968b: "Pentagon, pressgrupper og politikk [Pentagon, Pressure Groups, and Politics]", in: *Pax*, 7(1): 6–11.

NPG, 1968c: "Some Comments on Nonviolent Defense Research", in: Mohn, Reinhard (Ed.): *Civilian Defense* (Bielefeld: Bertelsmann Universitätsverlag): 155–164.

NPG, 1969a: "Hawaii og legenden om paradiset [Hawaii and the Myth of Paradise]", in: *Samtiden*, 78,8: 472–479.

NPG, 1969b: "The International Airline Network: A Test of the Zipf and Stouffer Hypotheses", in: *Papers, Peace Research Society (International)*, 11: 123–153.

NPG, 1970b: "Rank and Interaction: A General Theory with Some Applications to the International System", Proceedings of the IPRA Conference, IPRA Studies in Peace Research (Assen: van Gorcum): 1–21.

Høivik, Tord; NPG, 1970: "Structural Parameters of Graphs: A Theoretical Investigation", in: *Quality and Quantity*, 4,1: 193–209. Reprinted in: Blalock, Hubert M. (Ed.), 1975: *Quantitative Sociology* (New York: Academic Press): 203–224.

NPG, 1971: "Interaction Patterns in the Middle East", in: *Cooperation and Conflict*, 6,1: 15–30. Finnish Translation: Lähi-Idän vuorovaikutusknviot. In: Vesa, Unto (Ed.), 1971: *Sodat, kriisit ja rauhantuutkimus* (Tampere: TAPRI): 77–104.

NPG; Hartmann, Åke; Naustdalslid, Jon, 1971: "Mardøla-aksjonen [The Mardøla Demonstration]", in: *Tidsskrift for samfunnsforskning*, 12,3: 177–210.

NPG; Høivik, Tord, 1971: "Simulating Structural Parameters of Graphs: First Results", in: *Quality and Quantity*, 5,1: 224–227.

NPG, 1972: "Generaler og fotfolk i utakt [Generals and Rank and File Out of Step]", in: *Internasjonal Politikk*, 26,4B Supplement: 795–804.

NPG; Hellevik, Ottar, 1973: "The Common Market Decision in Norway: A Clash between Direct and Indirect Democracy", in: *Scandinavian Political Studies*, 8: 227–235.

NPG, 1974a: "Salary Equalization in a Research Institute", in: *Scandinavian Forest Economics*, 3–4: 29–31.

NPG, 1974b: "Time Differences and International Interaction", in: *Cooperation and Conflict*, 9,2: 35–51. Reprinted as Chap. 3 of this volume.

NPG; Høivik, Tord; Hellevik, Ottar, 1974: "Noen enkle modeller for valg og folkeavstemninger [Some Simple Models for Elections and Referenda]", in: *Tidsskrift for samfunnsforskning*, 15,3: 233–268.

Galtung, Johan; NPG, 1975: "Norge i verdenssamfunnet [Norway in the World Community]", in: Ramsøy, Natalie R.; Vaa, Mariken (Eds.): *Det Norske Samfunn*, 2nd edn, vol II (Oslo: Gyldendal): 742–811.

NPG, 1975: "Slow is Beautiful. The Stratification of Personal Mobility with Special Reference to International Aviation", in: *Acta Sociologica*, 18,1: 76–94.

NPG; Singer, J. David, 1975: "Distance and International War, 1816–1965", Proceedings of the IPRA Conference, IPRA Studies in Peace Research (Oslo: International Peace Research Association): 481–506.

Hellevik, Ottar; NPG; Ringdal, Kristen, 1975: "The Common Market Issue in Norway: A Conflict Between Center and Periphery", in: *Journal of Peace Research*, 12,1: 37–53.

NPG, 1976: "Hvordan og hvorfor Norge fikk Loran C [How and Why Norway got Loran-C]", in: *Internasjonal Politikk*, 34,4: 823–841.

NPG, 1977: "Towards a Multilateral Aviation Treaty", in: *Journal of Peace Research*, 14,3: 239–259.

NPG; Lodgaard, Sverre, 1977: "Norway—The Not so Reluctant Ally", in: *Cooperation and Conflict*, 12,4: 209–219. Originally published in Finnish in *Ulkopolitiikka*.

NPG; Høivik, Tord, 1978: "Best Interaction Models", in: *Quality and Quantity*, 12,4: 299–329. Reprinted in: Alker, Hayward (Ed.), 1979: *Mathematical Approaches to International Organizations* (Bucharest: Romanian Academy of Social and Political Sciences): 65–91.

Wilkes, Owen; NPG, 1978: "Optical Satellite Tracking: University Participation in Preparations for Space Warfare", in: *Journal of Peace Research*, 15,3: 205–225.

NPG, 1979: "Loran C og Polaris: Hvem visste hva og når fikk de vite det? [Loran-C and Polaris: Who Knew What and When Were They Informed?]", in: *Internasjonal Politikk*, 37,3: 405–420.

NPG, 1980: "A Salary System for Peace Research Institutes? Some PRIO Experiences", in: *International Peace Research Newsletter*, 18,2: 12–23.

NPG, 1980: "The Structure of Galtungism", in: Gleditsch, Nils Petter et al. (Eds.): *Johan Galtung. A Bibliography of His Scholarly and Popular Writings 1951–1980* (Oslo: PRIO): 64–81.

Wilkes, Owen; NPG, 1981: "Research on Intelligence or Intelligence as Research", in: Jahn, Egbert; Sakamoto, Yoshikazu (Eds.): Elements of World Instability: Armaments, Communication, Food, International Division of Labour. Proceedings of the IPRA Eight General Conference (Frankfurt am Main: Campus): 283–300. Abbreviated version in: Holm, Hans-Henrik; Rudeng, Erik (Eds.), 1980: *Social Science for What? Festschrift for Johan Galtung* (Oslo: Norwegian University Press): 170–181.

NPG; Bjerkholt, Olav; Cappelen, Ådne, 1983: "Conversion: Global, National, and Local Effects. A Case Study of Norway", in: *Cooperation and Conflict*, 18,3: 179–195. German version in: Wulf, Herbert (Ed.), 1983: *Abrüstung und Unterentwicklung. Berichte von Experten der Vereinten Nationen* (Reinbek: Rowohlt): 255–278. Slightly revised version as Conversion Effects: A Case Study of Norway. In: Dumas, Lloyd; Thee, Marek (Eds.), 1989: *Making Peace Possible: The Promise of Economic Conversion* (London: Pergamon): 231–249.

NPG; Bjerkholt, Olav; Cappelen, Ådne; Moum, Knut, 1983: "The Economic Effects of Conversion: A Case Study of Norway", in: Tuomi, Helena; Väyrynen, Raimo (Eds.): *Militarization and Arms Production* (London & New York: Croom Helm & St. Martin's): 225–258.

Cappelen, Ådne; NPG; Bjerkholt, Olav, 1984: "Military Spending and Economic Growth in the OECD Countries", in: *Journal of Peace Research*, 21,4: 361–373. Reprinted in: Hartley, Keith; Sandler, Todd (Eds.), 2002: *The Economics of Defence*. International Library of Critical Writings in Economics (Cheltenham: Elgar): 460–472.

NPG; Høgetveit, Einar, 1984: "Freedom of Information and National Security. A Comparative Study of Norway and United States", in: *Journal of Peace Research*, 21,1: 17–45.

NPG, 1985: "Arne Treholt – et politisk offer? [Arne Treholt—A Political Victim?]", in: *Ny Tid*, 27 June: 7.

NPG, 1986a: "Fredsforskning og politikk [Peace Research and Politics]", in: *Politica*, 18,3: 252–262.

NPG, 1986b: "The Strategic Significance of the Nordic Countries", in: *Current Research on Peace and Violence*, 9,1–2: 28–42. Abbreviated version as (1985) "Europe's Northern Region between the Superpowers", in: *Bulletin of Peace Proposals*, 16,4: 399–411. Revised version in Danish in: NPG; Møller, Bjørn; Wiberg, Håkan; Wæver, Ole, 1990: *Svaner på vildveje. Nordens sikkerhed mellem supermagtsflåder og europæisk opbrud* (Copenhagen: Vindrose): 28–56.

Sørdahl, Roger; NPG, 1986: "Kilder til sikkerhetspolitisk etterkrigshistorie [Sources to the History of Post-war Security Policy]", in: *Historisk tidsskrift*, 65,3: 273–317.

Skomsvold, Rolf; NPG; Cappelen, Ådne; Bjerkholt, Olav, 1987: "Regionaløkonomiske konsekvenser av nedrustning i Norge [Regional Economic Consequences of Disarmament in Norway]", in: *Sosiologi idag*, 17,3–4: 113–130.

Wilkes, Owen; NPG, 1987: "NAROL—An Early Attack Assessment System", in: *Intelligence and National Security*, 2,2: 331–335.

NPG, 1988a: "Etableringen av et internasjonalt tidsskrift: Journal of Peace Research [The Establishment of an International Journal: Journal of Peace Research]", in: Lundberg, Elizabeth (Ed.): *Internationell vetenskaplig publicering i Norden* (Copenhagen: Nordic Council of Ministers): 69–94.

NPG, 1988b: "Hemmelighold: Den britiske modellen [Secrecy: The British Model]", in: *Norsk statsvitenskapelig tidsskrift*, 4,2: 173–193.

NPG; Bjerkholt, Olav; Cappelen, Ådne, 1988: "Military R&D and Economic Growth in Industrialized Market Economies", in: Wallensteen, Peter (Ed.): *Peace Research: Achievements and Challenges* (Boulder, CO & London: Westview): 198–215.

NPG, 1989a: "Focus on: Journal of Peace Research", in: *Journal of Peace Research*, 26,1: 1–5.

NPG, 1989b: "Massemediene og sikkerhetspolitikken [The Mass Media and Security Policy]", in: Thomsen, Nils (Ed.): *Pressens Årbog* (Copenhagen & Fredrikstad: Reitzel & Institutt for Journalistikk, for Pressehistorisk Selskab): 37–46.

NPG, 1989c: "Oppbyggingen av et konsulentsystem for et samfunnsvitenskapelig tidsskrift [Building a Referee System for a Social Science Journal]", in: *NOP-nytt*, 15,4: 27–49.

NPG, 1990a: "The Development of Peace Research: An Editor's Perspective", in: Nobel, Jaap W. (Ed.): *The Coming of Age of Peace Research* (Groningen: Styx): 79–87.

NPG, 1990b: "Research on Arms Races", in: NPG; Njølstad, Olav (Eds.): *Arms Races—Technological and Political Dynamics* (London: SAGE): 1–14.

NPG, 1990c: "The Rise and Decline of the New Peace Movement", in: Kodama, Katsuya et al. (Eds.): *Towards a Comparative Analysis of Peace Movements* (Aldershot & Brookfield, VT: Dartmouth & Gower): 73–88.

NPG; Wolland, Steingrim, 1991: "Norway", in: D'Souza, Frances et al. (Eds.): *Information Freedom and Censorship. World Report 1991* (London: Library Association): 287–291.

Cappelen, Ådne; NPG; Bjerkholt, Olav, 1992: "Guns, Butter, and Growth: The Case of Norway", in: Chan, Steve; Mintz, Alex (Eds.): *Defense, Welfare, and Growth* (London & New York: Routledge): 61–80.

NPG, 1992b: "Defense without Threat? The Future of Norwegian Military Spending", in: *Cooperation and Conflict*, 27,4: 397–413.

NPG, 1992c: "Democracy and Peace", in: *Journal of Peace Research*, 29,4: 369–376. Reprinted as Chap. 4 of this volume. Revised version (1993) as "Democracy and Peace: Good News for Human Rights Advocates", in: Gomien, Donna (Ed.): *Broadening the Frontiers of Human Rights: Essays in Honour of Asbjørn Eide* (Oslo & Oxford: Scandinavian University Press & Oxford University Press): 283–306.

NPG; Wiberg, Håkan; Smith, Dan, 1992: "The Nordic Countries: Peace Dividend or Security Dilemma", in: *Cooperation and Conflict*, 27,4: 323–347.

NPG, 1993: "The Most-Cited Articles in JPR", in: *Journal of Peace Research*, 30,4: 445–449.

NPG; Agøy, Nils Ivar, 1993: "Norway: Toward Full Freedom of Choice", in: Moskos, Charles C.; Chambers, John Whiteclay (Eds.): *The New Conscientious Objection: From Sacred to Secular Resistance* (New York: Oxford University Press): 114–126.

NPG, 1994a: "Conversion and the Environment", in: Käkönen, Jyrki (Ed.): *Green Security or Militarized Environment* (Aldershot & Brookfield, VT: Dartmouth): 131–154.

NPG, 1994b: "Treholt-litteraturen: En studie i polarisering [The Treholt Literature: A Study in Polarization]", in: *Internasjonal Politikk*, 52,2: 275–287.

NPG, 1995a: "35 Major Wars? A Brief Comment on Mueller", in: *Journal of Conflict Resolution*, 39,3: 584–587.

NPG, 1995b: "Democracy and the Future of European Peace", in: *European Journal of International Relations*, 1,4: 539–571. Shorter version as (1995) "Demokratie, Krieg und die Zukunft Europas", in: *Welt Trends*, 10: 109–127.

NPG, 1995c: "Freedom of Information and National Security: A Comparative Perspective", in: Nagel, Stuart S. (Ed.): *Research in Law and Policy Studies*, vol. 4 (Greenwich, CT & London: JAI Press): 25–43.

NPG, 1995d: "Geography, Democracy, and Peace", in: *International Interactions*, 20,4: 297–323.

NPG, 1995e: "The Treholt Case—A Review of the Literature", in: *Intelligence and National Security*, 10,3: 529–538. Reprinted as Chap. 5 of this volume.

Cappelen, Ådne; NPG; Bjerkholt, Olav, 1996: "The Peace Dividend in Norway: Domestic or International?", in: NPG et al. (Eds.) *The Peace Dividend. Contributions to Economic Analysis.* 235 (Amsterdam: North-Holland): 275–303.

NPG, 1996a: "The APSR Hall of Fame: A Comment", in: *PS: Political Science and Politics*, 29,4: 637–638.

NPG, 1996b: "Democracy and Democratization: Not Quite the State of the Art", in: *Security Dialogue*, 27,3: 349–354.

NPG, 1996c: "Det nye sikkerhetsbildet: Mot en demokratisk og fredelig verden? [The New Security Picture: Towards a Democratic and Peaceful World?]", in: *Internasjonal Politikk*, 54,3: 291–310. In revised form as Chap. 1 in: NPG et al. (Eds.), 1998: *Det nye sikkerhetsbildet* (Trondheim: Tapir): 9–28.

Ellingsen, Tanja; NPG, 1997: "Democracy and Armed Conflict in the Third World", in: Smith, Dan; Volden, Ketil (Eds.): *Causes of Conflict in the Third World* (Oslo: North-South Coalition & PRIO): 69–81.

NPG, 1997: "Environmental Conflict and the Democratic Peace", in: NPG et al. (Eds.): *Conflict and the Environment* (Dordrecht: Kluwer Academic): 91–106.

NPG; Hegre, Håvard, 1997: "Peace and Democracy: Three Levels of Analysis", in: *Journal of Conflict Resolution*, 41,2: 283–310.

NPG, 1998a: "Armed Conflict and the Environment. A Critique of the Literature", in: *Journal of Peace Research*, 35,3: 381–400. Reprinted in: Diehl, Paul F. (Ed.), 2005: *War*, vol. III (London: Sage): 252–274; in: Mitchell, Ronald B. (Ed.), 2008: *International Environmental Politics,* vol. IV (London: Sage): 237–258; in: Matthew, Richard A. (Ed.), 2014: *Environmental Security,* vol. II: *Environmental Change, National Security and the Conflict Cycle* (London: Sage): 175–198; in: Stern, David I.; Jotzo, Frank; Dobes, Leo (Eds.), 2014: *Climate Change and the World Economy.* International Library of Critical Writings in Economics (Cheltenham: Elgar): 641–660, and as Chap. 6 of this volume. Revised and shortened version in: Diehl, Paul F.; NPG (Eds.), 2001: *Environmental Conflict* (Boulder, CO: Westview): 251–272.

NPG, 1998b: "Fred og demokrati [Peace and Democracy]", in: Midgaard, Knut; Rasch, Bjørn Erik (Eds.) *Demokrati – vilkår og virkninger* (Oslo & Bergen: Fagbokforlaget): 299–316. Revised version by NPG; Håvard Hegre in the second edition (2004), 293–322.

McLaughlin, Sara; Gates, Scott; Hegre, Håvard; Gissinger, Ranveig; NPG, 1998: "Timing the Changes in Political Structures: A New Polity Database", in: *Journal of Conflict Resolution*, 42,2: 231–242.

Gissinger, Ranveig; NPG, 1999: "Globalization and Conflict: Welfare, Distribution, and Political Unrest", in: *Journal of World-Systems Research*, 5,2: 327–365.

NPG, 1999a: "Do Open Windows Encourage Conflict?", in: *Statsvetenskaplig tidskrift*, 102,3: 333–349.

NPG, 1999b: "Freedom of Expression, Freedom of Information, and National Security: The Case of Norway", in: Coliver, Sandra et al. (Eds.): *Secrecy and Liberty: National Security, Freedom of Expression and Access to Information* (Haag: Nijhoff): 361–388.

NPG, 1999c: "Peace and Democracy", in: Kurtz, Lester (Ed.): *Encyclopedia of Violence, Peace, and Conflict*, vol. 2 (San Diego, CA: Academic Press): 643–652. Revised version in second edition (2008), 1430–1437.

Toset, Hans Petter Wollebæk; NPG; Hegre, Håvard, 2000: "Shared Rivers and Interstate Conflict", in: *Political Geography*, 19,8: 971–996. Revised version as: NPG; Hegre, Håvard; Toset, Hans Petter Wollebæk, 2007: "Conflicts in Shared River Basins", in: Grover, Velma (Ed.): *Water: A Source of Conflict or Cooperation?* (Enfield, NH: Science Publishers): 39–66.

Gates, Scott; Hegre, Håvard; NPG, 2001: "Democracy and Civil Conflict After the Cold War", in: Berg-Schlosser, Dirk; Vetik, Raivo (Eds.): *Perspectives on Democratic Consolidation in Central and Eastern Europe* (New York: Columbia University Press, for East European Monographs): 185–194.

NPG, 2001a: "Environmental Change, Security, and Conflict", in: Crocker, Chester; Hampson, Fen Osler; Aall, Pamela (Eds.): *Turbulent Peace: The Challenges of Managing International Conflict* (Washington, DC: United States Institute of Peace Press): 53–68. Revised version in: Crocker, Chester; Hampson, Fen Osler; Aall, Pamela (Eds.), 2007: *Leashing the Dogs of War: Conflict Management in a Divided World* (Washington, DC: United States Institute of Peace Press): 177–195. French version: Changements environne-mentaux, sécurité et conflits. In: Hallegatte, Stéphane; Ambrosi, Philippe (Eds.), 2007: "Environnement, changement climatique et sécurité. Questions scientif-iques et enjeux opérationnels", Special issue of *Les Cahiers de la Sécurité*, 63,4: 121–156. Spanish version: Cambio medioambiental, seguridad y conflicto. In: Sanahuja, José Antonio (Ed.), 2012: *Construcción de la paz, seguridad y desarrollo. Visiones, políticas y actores* (Madrid: Editorial Complutense): 99–125. Revised version: Climate Change, Environmental Stress, and Conflict. In: Crocker, Chester; Hampson, Fen Osler; Aall, Pamela (Eds.), 2015: *Conflict Management and Global Governance in an Age of Awakening* (Washington, DC: United States Institute of Peace Press): 147–168.

NPG, 2001b: "Mot et utvidet sikkerhetsbegrep? [Toward an Extended Concept of Security?]", in: Hovi, Jon; Malnes, Raino (Eds.): *Normer og makt: Innføring i internasjonal politikk* (Oslo: Abstrakt): 95–114.

NPG, 2001c: "Resource and Environmental Conflict: The State-of-the-Art", in: Petzold-Bradley, Eileen; Carius, Alexander; Vincze, Arpád (Eds.): *Responding to Environmental Conflicts: Implications for Theory and Practice*. Nato Science Series 2. Environmental Security, 78 (Dordrecht: Kluwer): 53–66.

Hegre, Håvard; Ellingsen, Tanja; Gates, Scott; NPG, 2001: "Toward a Democratic Civil Peace? Democracy, Political Change, and Civil War, 1816–1992", in: *American Political Science Review*, 95,1: 17–33. Reprinted in: Diehl, Paul F. (Ed.), 2005: *War*. Library of International Relations, vol. V (London: Sage): 165–193.

NPG, 2002a: "Borgerkrig – vår tids landeplage? [Civil War—The Scourge of Our Time?]", in: *Økonomisk Forum*, 56,1: 4–7.

NPG, 2002b: "Double-Blind but More Transparent", in: *Journal of Peace Research*, 39,3: 259–262. Reprinted as Chap. 7 of this volume.

NPG, 2002c: "Fra krig om olje til krig om vann? [From Oil Wars to Water Wars?]", in: *P2-akademiet,* Book Y (Oslo: NRK Fakta): 34–45.

NPG; Sverdrup, Bjørn Otto, 2002: "Democracy and the Environment", in: Page, Edward A.; Redclift, Michael (Eds.): *Human Security and the Environment: International Comparisons* (Cheltenham: Edward Elgar): 45–70.

NPG; Urdal, Henrik, 2002: "Ecoviolence? Links Between Population Growth, Environmental Scarcity, and Violent Conflict in Thomas Homer-Dixon's Work", in: *Journal of International Affairs*, 56,1: 283–302.

NPG; Wallensteen, Peter; Eriksson, Mikael; Sollenberg, Margareta; Strand, Håvard, 2002: "Armed Conflict 1946–2001: A New Dataset", in: *Journal of Peace Research*, 39,5: 615–637.

Furlong, Kathryn; NPG, 2003: "The Boundary Dataset", in: *Conflict Management and Peace Science*, 20,1: 93–117.

NPG, 2003a: "Arvelig belastet [The Burden of Heritage]", in: Eek, Øystein (Ed.): *Gode Gamle Katta* (Oslo: Oslo katedralskole): 178–179.

NPG, 2003b: "Environmental Conflict: Neomalthusians vs. Cornucopians", in: Brauch, Hans Günter et al. (Eds.): *Security and the Environment in the Mediterranean: Conceptualising Security and Environmental Conflicts* (Berlin: Springer): 477–485.

Gleditsch, Nils Petter, 2003c: "Mappa mi (My file)", in: *Dagbladet*, 20 December, www.dagbladet.no/kultur/2003/12/20/386616.html. Expanded version as 'Mappene våre' [Our files]. In: Ofstad, Bjørg; Bjerkholt, Olav; Skrede, Kari; Hylland, Aanund (Eds.), 2009: *Rettferd og politikk. Festskrift til Hilde Bojer* (Oslo: Emilia): 351–356.

NPG; Metelits, Claire, 2003: "The Replication Debate", in: *International Studies Perspectives*, 4,1: 72–79.

NPG; Metelits, Claire; Strand, Håvard, 2003: "Posting Your Data: Will You Be Scooped or Will You Be Famous?", in: *International Studies Perspectives*, 4,1: 89–97.

Hegre, Håvard; Gissinger, Ranveig; NPG, 2003: "Globalization and Internal Conflict", in: Schneider, Gerald; Barbieri, Katherine; NPG (Eds.): *Globalization and Armed Conflict* (Lanham, MD: Rowman & Littlefield): 251–276.

Ravlo, Hilde; NPG; Dorussen, Han, 2003: "Colonial War and the Democratic Peace", in: *Journal of Conflict Resolution*, 47,4: 520–548.

Schneider, Gerald; Barbieri, Katherine; NPG, 2003: "Does Globalization Contribute to Peace? A Critical Survey of the Literature", in: Schneider, Gerald; Barbieri, Katherine; NPG (Eds.): *Globalization and Armed Conflict* (Lanham, MD: Rowman & Littlefield): 3–30.

NPG, 2004: "Peace Research and International Relations in Scandinavia: From Enduring Rivalry to Stable Peace?", in: Guzzini, Stefano; Jung, Dietrich (Eds.): *Copenhagen Peace Research: Conceptual Innovation and Contemporary Security Analysis. Essays in Honour of Håkan Wiberg* (London: Routledge): 15–26.

NPG; Hegre, Håvard, 2004: "En globalisert verden – økt kaos eller varig fred? [A Globalized World—Increased Chaos or Lasting Peace?]", in: Snoen, Jan Arild (Ed.): *Åpen verden. Et forsvar for globaliseringen* (Oslo: Civita): 250–263.

Gilmore, Elisabeth; NPG; Lujala, Päivi; Rød, Jan Ketil, 2005: "Conflict Diamonds: A New Dataset", in: *Conflict Management and Peace Science*, 22,3: 257–292.

Lacina, Bethany; NPG, 2005: "Monitoring Trends in Global Combat: A New Dataset of Battle Deaths", in: *European Journal of Population*, 21,2–3: 145–166. Reprinted as Chap. 6 in: Brunborg, Helge; Tabeau, Ewa; Urdal, Henrik (Eds.). 2006: *The Demography of Armed Conflict* (Dordrecht: Springer): 131–151.

Lujala, Päivi; NPG; Gilmore, Elisabeth, 2005: "A Diamond Curse? Civil War and a Lootable Resource", in: *Journal of Conflict Resolution*, 49,4: 538–562.

Sørli, Mirjam E.; NPG; Strand, Håvard, 2005: "Why Is There So Much Conflict in the Middle East?", in: *Journal of Conflict Resolution*, 49,1: 141–165.

Buhaug, Halvard; NPG, 2006: "The Death of Distance? The Globalization of Armed Conflict", in: Kahler, Miles; Walter, Barbara (Eds.): *Territoriality and Conflict in an Era of Globalization* (Cambridge: Cambridge University Press): 187–216.

Furlong, Kathryn; NPG; Hegre, Håvard, 2006: "Geographic Opportunity and Neomalthusian Willingness: Boundaries, Shared Rivers, and Conflict", in: *International Interactions*, 32,1: 79–108.

NPG; Furlong, Kathryn; Hegre, Håvard; Lacina, Bethany; Owen, Taylor, 2006: "Conflicts over Shared Rivers: Resource Scarcity or Fuzzy Boundaries?", in: *Political Geography*, 25,4: 361–382.

Lacina, Bethany; NPG; Russett, Bruce, 2006: "The Declining Risk of Death in Battle", in: *International Studies Quarterly*, 50,3: 673–680.

Binningsbø, Helga Malmin; de Soysa, Indra; NPG, 2007: "Green Giant or Straw Man? Environmental Pressure and Civil Conflict, 1961–99", in: *Population and Environment*, 28,6: 337–353.

NPG, 2007a: "Incentives to Publish?", in: *European Political Science*, 6,2: 185–191.

NPG, 2007b: "Tellekantenes fundament [The Basis for Counting Publications]", in: *Morgenbladet*, 30 March: 26.

NPG, 2007c: "Årsaker til krig [Causes of War]", in: Hovi, Jon; Malnes, Raino (Eds.): *Anarki, makt og normer. Innføring i internasjonal politikk*. Second edition (Oslo: Abstrakt): 166–184. Audio version, Norsk lyd- og blindeskriftbibliotek (2007). Revised version with Halvard Buhaug (2011), 167–190.

Nordås, Ragnhild; NPG, 2007: "Climate Change and Conflict", in: *Political Geography*, 26,6: 627–638.

Buhaug, Halvard; NPG; Theisen, Ole Magnus, 2008: *Implications of Climate Change for Armed Conflict*. Paper prepared for the Social Dimensions of Climate Change program (Washington, DC: World Bank, Social Development Department). http://siteresources.worldbank.org/INTRANETSOCIALDEVELOPMENT/Resources/SDCCWorkingPaper_Conflict.pdf. Shorter version as Implications of climate change for armed conflict. Chapter 3 in: Mearns, Robin; Norton, Andy (Eds.), 2010: *Social Dimensions of Climate Change: Equity and Vulnerability in a Warming World. New Frontiers of Social Policy* (Washington, DC: World Bank): 75–101 and as Climate change and armed conflict. Chapter 9 in: Brown, Graham; Langer, Arnim (Eds.), 2012: *Elgar Companion to Civil War and Fragile States* (London: Edward Elgar): 125–138.

NPG, 2008a: "The Liberal Moment Fifteen Years On. Presidential address, International Studies Association", in: *International Studies Quarterly*, 52,4: 691–712. Reprinted as Chap. 8 of this volume.

NPG, 2008b: "Klimaendringer og sikkerhet [Climate Change and Security]", in: *FN-magasinet*, 2,1: 9–11.

NPG; Urdal, Henrik, 2008: "Om topptidsskrifter i internasjonal politikk [On Top Journals in International Relations]", in: *Internasjonal Politikk*, 66,4: 699–701.

Rustad, Siri Camilla Aas; Rød, Jan Ketil; Larsen, Wenche; NPG, 2008: "Foliage and Fighting: Forest Resources and the Onset, Duration, and Location of Civil War", in: *Political Geography*, 27,7: 761–782.

Urdal, Henrik; Theisen, Ole Magnus; NPG; Buhaug, Halvard, 2010: "Klimakriger? En vurdering av det faglige grunnlaget [Climate Wars? An Evaluation of the Academic Basis]", in: *Norsk Statsvitenskapelig Tidsskrift*, 26,4: 297–320.

NPG; Hegre, Håvard; Strand, Håvard, 2009: "Democracy and Civil War", in: Midlarsky, Manus (Ed.): *Handbook of War Studies III* (Ann Arbor, MI: University of Michigan Press): 155–192 + refs. 301ff.

NPG; Nordås, Ragnhild, 2009: "Climate Change and Conflict: A Critical Overview", in: *Die Friedens-Warte*, 84,2: 11–28. Revised version in: Hartard, Susanne; Liebert, Wolfgang (Eds.), 2015: *Competition and Conflicts on Resource Use* (Heidelberg: Springer): 21–38.

NPG, 2010: "Introductory Essay", in: Young, Nigel J. (Ed.): *The Oxford International Encyclopedia of Peace* (Oxford: Oxford University Press): xxxiii–xxxvi.

NPG; Theisen, Ole Magnus, 2010: "Resources, the Environment, and Conflict", in: Cavelty, Myriam Dunn; Mauer, Victor (Eds.): *Routledge Handbook of Security Studies* (London: Routledge): 221–231. Revised version, second edition (2015), in press.

Schneider, Gerald; NPG, 2010: "A Capitalist Peace?", in: *International Interactions*, 36,2: 107–114. Revised version as "The Capitalist Peace: The Origins and Prospects of a Liberal Idea", in: Schneider, Gerald; NPG (Eds.), 2013: *Assessing the Capitalist Peace* (London: Routledge): 1–9.

Schneider, Gerald; NPG; Carey, Sabine C., 2010: "Exploring the Past, Anticipating the Future: A Symposium", in: *International Studies Review*, 12,1: 1–7.

Schneider, Gerald; Carey, Sabine C.; NPG, 2011: "Forecasting in International Relations: One Quest, Three Approaches", in: *Conflict Management and Peace Science*, 28,1: 5–14.

NPG; Listhaug, Ola, 2011: "Foreword", in: Jakobsen, Tor G. (Ed.): *War. An Introduction to Theories and Research on Collective Violence* (Hauppauge, NY: Nova): vii.

Bernauer, Thomas; Böhmelt, Tobias; Buhaug, Halvard; NPG; Weibust, Eivind Berg; Wischnath, Gerdis, 2012: "Intrastate Water-Related Conflict and Cooperation: A New Event-Dataset (WARRIC)", in: *International Interactions*, 38,4: 529–545.

Bernauer, Thomas; NPG, 2012: "Introduction to the Special Issue on Events Data in Conflict", in: *International Interactions*, 38,4: 375–381.

Brochmann, Marit; NPG, 2012: "Shared Rivers and Conflict—A Reconsideration", in: *Political Geography*, 31,8: 519–527.

Brochmann, Marit; Rød, Jan Ketil; NPG, 2012: "International Borders and Conflict Revisited", in: *Conflict Management and Peace Science*, 29,2: 170–194.

Cappelen, Ådne; NPG, 2012: "En fredsgevinst for Norge – eller fortsatt opprustning? [A Peace Dividend for Norway—Or Continued Rearmament?]", in: *Samfunnsøkonomen*, 26,6: 26–31.

NPG, 2012c: "Aldri for sent å være pessimist? [Never Too Late To Be a Pessimist?]", in: Larsen, Sverre Røed; Hjort-Larsen, Anne (Eds.): *I strid for fred. Fredskontoret 1962–1972* (Oslo: Kolofon): 179–185.

NPG, 2012d: "Blir det mer krig i verden? [More War in the World?]", in: Krøvel, Roy; Orgeret, Kristin Skare (Eds.): *Historier om verden. Utvikling og miljø i globalt perspektiv* (Kristiansand: IJ-forlaget): 60–61.

NPG, 2012e: "Whither the Weather? Climate Change and Conflict", in: *Journal of Peace Research*, 49,1: 4–9. Reprinted as Chap. 9 of this volume.

NPG, 2012f: "Open Access in International Relations: A Symposium", in: *International Studies Perspective*, 13,3: 211–215.

NPG; Urdal, Henrik, 2012: "Står barometeret på storm? Ny forskning gir lite støtte til tanken om klimakonflikter [A Storm Warning on the Barometer? New Research Provides Little Support for the Idea of Climate Conflicts]", in: *Klima*, 13,1: 30–31. Norwegian version and English translation at www.cicero.uio.no.

NPG, 2013a: "Arthur Westing: A Personal Memoir". Preface to Brauch, Hans Günter (Ed.): *Arthur H Westing—From Environmental to Comprehensive Security*. SpringerBriefs on Pioneers in Science and Practice, 1 (Heidelberg: Springer): xi–xiii.

NPG, 2013b: "The Decline of War—The Main Issues", in: NPG (Ed.): "The Forum: The Decline of War", in: *International Studies Review*, 15,3: 397–399. Reprinted as Chap. 10 of this volume.

Lacina, Bethany Ann; NPG, 2013: "The Waning of War is Real: A Reply to Gohdes and Price", in: *Journal of Conflict Resolution*, 57,6: 1109–1127.

Nordås, Ragnhild; NPG, 2013: "The IPCC, Human Security, and the Climate-Conflict Nexus", in: Redclift, Michael; Grasso, Marco (Eds.): *Handbook on Climate Change and Human Security* (London: Elgar): 67–88.

Østby, Gudrun; Strand, Håvard; NPG; Nordås, Ragnhild, 2013: Gender Gap or Gender Bias in Peace Research? Publication Patterns for Journal of Peace Research, 1983–2008", in: *International Studies Perspectives*, 14,4: 493–506.

Theisen, Ole Magnus; NPG; Buhaug, Halvard, 2013: "Is Climate Change a Driver of Armed Conflict?", in: *Climatic Change*, 117,3: 613–625.

Buhaug, Halvard; Nordkvelle, Jonas; Bernauer, Thomas; Böhmelt, Tobias; Brzoska, Michael; Busby, Josh W.; Ciccone, Antonio; Fjelde, Hanne; Gartzke, Erik; NPG; Goldstone, Jack A.; Hegre, Håvard; Holtermann, Helge; Koubi, Vally; Link, Jasmin S.A.; Link, Peter Michael; Lujala, Päivi; O'Loughlin, John; Raleigh, Clionadh; Scheffran, Jürgen; Schilling, Janpeter; Smith, Todd G.; Theisen, Ole Magnus; Tol, Richard S.J.; Urdal, Henrik; von Uexkull, Nina, 2014: "One Effect to Rule Them All? A Comment on Climate and Conflict", in: *Climatic Change*, 127,3–4: 391–397.

Bussmann, Margit; Dorussen, Han; NPG, 2014: "Against All Odds: 2013 Richardson Award to Mats Hammarström and Peter Wallensteen", in: *Peace Economics, Peace Science and Public Policy*, 20,2: 235–243.

NPG, 2014a: "Klimaendringer og krig. FNs klimapanels rapport spår ikke flere kriger, mener fredsforsker [Climate Change and War. The IPCC Does Not Predict More War, Peace Researcher Argues]", in: *Aftenposten*, 11 April. English translation at http://blogs.prio.org/2014/04/climate-change-and-war/. Reprinted as Chap. 11 of this volume.

NPG, 2014b: "Overvåking under kontroll [Surveillance under Control]", in: Berg Lene (Ed.): *Gompen og andre beretninger om overvåking i Norge 1948–89* (Oslo: URO/KORO): 7–16. Slightly revised and updated version in *Nytt Norsk Tidsskrift*, 31,5: 441–451. English translation at http://blogs.prio.org/2015/01/surveillance-under-control/#more-1452.

NPG, 2014c: "Will Climate Change Reverse the Trend Towards Peace?", in: Schneckener, Ulrich; von Scheliha, Arnulf; Lienkamp, Andreas; Klagge, Britta (Eds.) *Wettstreit um Ressourcen. Konflikte um Klima, Wasser und Boden* (Berlin: Oekom): 49–60.

NPG; Nordås, Ragnhild, 2014: "Conflicting Messages? The IPCC on Conflict and Human Security", in: *Political Geography*, 43(November): 82–90.

NPG; Nordkvelle, Jonas; Strand, Håvard, 2014: "Peace Research—Just the Study of War?", in: *Journal of Peace Research*, 51,2: 145–158.

Urdal, Henrik; Østby, Gudrun; NPG, 2014: "Journal of Peace Research", in: *Peace Review*, 26,4: 500–504.

Böhmelt, Tobias; Bernauer, Thomas; Buhaug, Halvard; NPG; Tribaldos, Theresa; Wischnath, Gerdis, 2014: "Demand, Supply, and Restraint: Determinants of Domestic Water Conflict and Cooperation", in: *Global Environmental Change*, 29(November): 337–348.

NPG; Melander, Erik; Urdal, Henrik, 2015: "Introduction: Patterns of Armed Conflict Since 1945", in: Mason, David; Mitchell, Sara McLaughlin (Eds.): *What Do We Know About Civil War?* (Lanham, MD: Rowman & Littlefield, in press).

2.4 Selected Papers and Reports

NPG, 1968d: *The Structure of the International Airline Network*. Thesis for the mag.art. degree, University of Oslo.

Wilkes, Owen; NPG, 1979: *Intelligence Stations in Norway: Their Number, Location, Function, and Legality* (Oslo: PRIO).

NPG (Ed.), 1981a: *Dommen over Gleditsch & Wilkes. Fire kritiske innlegg* [The Sentence on Gleditsch and Wilkes. Four Critical Comments] (Oslo: PRIO).

NPG (Ed.), 1981b: *Forskning eller spionasje? Rapport om straffesaken i Oslo Byrett i mai 1981* [Research or Espionage? Report on the Criminal Trial at the Oslo City Court in May 1981] (Oslo: PRIO).

NPG (Ed.), 1981c: *The Oslo Rabbit Trial. A Record of the 'National Security Trial' against Owen Wilkes and NPG in the Oslo Town Court, May 1981* (Oslo: Solidarity Campaign for Gleditsch and Wilkes).

NPG, 1982: Annen runde. Høyesteretts behandling av straffesaken mot Gleditsch og Wilkes (Oslo: PRIO). [English version: Round Two. The Norwegian Supreme Court vs Gleditsch & Wilkes, February 1982 (Oslo: PRIO).]

NPG, 2002d: *The Future of Armed Conflict* (Ramat Gan: Bar-Ilan University, Begin-Sadat Center for Strategic Studies).

NPG; Christiansen, Lene S.; Hegre, Håvard, 2007: *Democratic Jihad? Military Intervention and Democracy*. Post-conflict Transitions Working Paper 15, WSPS 4242 (Washington, DC: Development Research Group, World Bank).

NPG; Nordås, Ragnhild; Salehyan, Idean, 2007: *Climate Change, Migration, and Conflict*. Coping with Crisis Working Paper Series (New York: International Peace Academy).

Rolseth, Amund; Theisen, Ole Magnus; NPG, 2014: "Violence Against Civilians 1900–87: Regime Type, Climate Change, and the Severity of Democide", Paper Presented at the Biannual Conference of the International Network of Genocide Scholars, Cape Town, 4–7 December.

Strand, Håvard; Nordkvelle, Jonas; Gleditsch, Nils Petter, 2014: "Posting Your Data: Will You Remain Famous?", Paper Presented at the 55th Annual Convention of the International Studies Association, Toronto, 26–29 March.

Part II
Key Texts by Nils Petter Gleditsch

Chapter 3
Time Differences and International Interaction

Physical distance appears to act as a restraint on interaction at all levels of social organization.[1] However, there is one specific problem connected with high-speed interaction over great distance in the international system—that of *time differences*.[2] In international travel the 'jet lag' causes fatigue and related phenomena. (The problem of a sudden *change of climate* associated with rapid North–South movements has not been studied to the same degree, but appears to be less serious.) In attempting to circumvent these unpleasant effects by interacting through telecommunication (moving information rather than moving people), one runs into a related problem—that of non-overlapping office hours. Informal data from several organizations with international activities are cited as examples of how these problems are dealt with. Technological and social 'solutions' to the problem of time differences are discussed. Several of these raise new problems, among them the possibility of an emerging 'time imperialism'—with dominant nations, organizations, and individuals imposing their own time cycles on their dependent individuals and groups—seems particularly ominous.

---

[1]This article was originally published in *Cooperation and Conflict* 9(1): 35–51, 1974.

[2]This article is the result of work done over a long time with many interruptions. Most of the data on time zones were collected while the author was a research associate of the Dimensionality of Nations Project, University of Hawaii, in 1969. The rest of the work was done at the International Peace Research Institute, Oslo and the article can be identified as PRIO publication no. 21–23. Previous versions have been presented to the Nordic conference in peace research, Fagerfjell, Norway, February 1972; a PRIO seminar, May 1972; and the IX Congress of the International Political Science Association, August 1973. I am grateful to Jon Naustdalslid for research assistance and to various professional colleagues for comments, particularly Johan Galtung, Johan Jørgen Holst, Tord Høivik, Arden Johnson, and Robert Klitgaard. I am also grateful to various people in business and government in Oslo for giving of their time to discuss these problems. Economic support has been provided by the Norwegian Council for Research in Science and the Humanities (NAVF) and the Norwegian Research Council for Conflict and Peace (RKF).— Postscript 2014: Two of the people who provided information on handling time differences in politics and business, were former Norwegian prime minister Einar Gerhardsen and Jan P. Syse, then a senior executive in Wilhelmsen's shipping line, and later also a prime minister. Why they were not thanked by name, I can no longer remember.

© The Author(s) 2015
N.P. Gleditsch, *Nils Petter Gleditsch: Pioneer in the Analysis of War and Peace*, SpringerBriefs on Pioneers in Science and Practice 29, DOI 10.1007/978-3-319-03820-9_3

3.1 Introduction

To winter sports enthusiasts in Norway, the XI Winter Olympic Games in Sapporo, Japan in February 1972 provided a free introduction to the 'Brave New World' in one interesting respect: Televising of the events started at 5:25 in the morning. This is interesting for two reasons: First of all, it was the first time since the Second World War that the Winter Olympic Games had been held outside Europe or the US —yet there was no question of European viewers not getting their full share of 'instant' news. 10 years earlier this could not have been done—15 years earlier the whole idea would still have been science fiction. Secondly, the transmissions could have been instantaneous but, in fact, were not. This was not because of any technical limitation of satellite communication; it must have been a conscious decision on the part of the Norwegian broadcasting corporation (NRK) that 5:25 was the earliest time one could decently shake Norwegian viewers out of bed. (For some reason, radio transmissions started at 5 am.) The reports were seen or heard by hundreds of thousands of Norwegians—according to a poll, 21.1 % of the population heard at least one early morning radio report and 33 % saw at least one TV program between 5:25 and 6:00.[3] For all these people, the NRK decision determined their daily sleep cycle for at least a day, in many cases for the best part of 2 weeks. A technological breakthrough led to a temporary change of life-style.

Just as remarkable as these two points are in themselves is the fact that all this occasioned very little comment. It was not, of course, the first time that a major sports event had been televised world-wide. For that matter, Norway had only had television for a little over a decade. But adjustment to the technological breakthrough had been so rapid that the changes in life-style that they required were hardly remarkable any more.

This example may serve as an introduction to the more general problem: What are some of the consequences of a rapid increase in the speed of communication? These consequences are often discussed under the heading of 'the shrinking globe' or 'the decreasing significance of distance in the international system'. We turn first to an examination of the concept of distance.

3.2 Geographical Distance: Horizontal and Vertical

At all levels of social organization, *physical distance* has a restraining impact on interaction. In a cafeteria, you may more easily—everything else being equal—fall into conversation with someone who shares your table than with someone across the room. In an apartment building, you will more easily get to know those who pass by your door on their way in or out (Festinger et al. 1949). That a similar relationship holds for the international system should come as no great surprise.

---

[3] According to a survey carried out by the Central Bureau of Statistics (1972).

In recent years, it has become fashionable to proclaim the shrinking world and the decreasing importance of distance in the international system. However, contrary to common thinking (and my own initial expectations), I found in a previous study that distance had *increased* its correlation with one form of international interaction (scheduled international flights) over the period 1930–65 (Gleditsch 1969).

Clearly, *geographical distance* itself is not the mechanism at work. *Straight-line distance* does not necessarily equal *functional distance*. First, the actual impediment to interaction may be *time* or economic *cost* and these may depend on a route structure in an existing interaction network or on physical factors. For pedestrians in a city, city-block distance is a more realistic measure of travel time than bee-line distance. For international air travelers, the belated introduction in 1967 of an air link across the Soviet Union cut travel time between Europe and Japan by as much as 25 %. But distance, and demand on facilities in turn, influence the route structure. Pedestrian passageways can be constructed through buildings in extremely busy sections of a town. When in 1965 Western Samoa was not linked to Australia except via American Samoa or the Fiji Islands, it was presumably because the demand was not heavy enough to justify a direct link across 2,847 miles.

Quite apart from the problem of defining and measuring functional distance in any social system, there is a specific peculiarity about distance in the international system: This is the basic distinction between *vertical distance*, or North–South distance, and *horizontal distance*, or East–West distance. Travelling in an East–West direction one has to overcome a difference in *local time*. In the North–South direction the difference in local climate is a corresponding hurdle. Table 3.1 spells out in detail some salient characteristics of the two.

The impact of vertical distance on international interaction will not be extensively discussed here. This problem occurs only when persons or goods are moved, not with the movement of information. In some cases it can be quite serious. A sudden change in climate (temperature, humidity) or in vegetation can have a marked physical effect on general well-being or specific diseases (such as allergy

Table 3.1 The two components of distance in the international system

	North/South	East/West
Direction	Vertical	Horizontal
Climate	Dissimilar	Similar
Local time	Similar	Dissimilar
Functional distance	Curved, continuous	Monotonic, stepwise
Creates problems in moving *persons* (travel)	Yes	Yes
Creates problems in moving *goods* (trade)	Yes	No
Creates problems in moving information (communication)	No	Yes
Problem is aggravated as functional distance decreases	Yes	Yes
Problem comes into existence only when the speed of movement is very high	No	Yes

conditions). For the community which receives the traveler, it also increases the risk of spreading epidemics. While a Near Eastern cholera epidemic a few decades ago could only have spread to Scandinavia via the intermediate European countries (and probably would have stopped or have been stopped on its way), it can spread today via, e.g., tourist charter flights. Since a flight is completed much faster than the incubation period for a disease, isolating the infected traveler at the point of destination is no longer a viable solution.

However, there are other effective measures for the prevention of epidemics, e.g. mass inoculation. Also, the increased speed of communication has been accompanied by an improvement in medical skills and (for the developed world, at least) an increase in the general health level to a point where epidemics are no longer so serious. More drastic and specific countermeasures are also available, such as disinfecting the planes and the requirement of specific vaccination for travel to certain countries (Leuschner 1965). Vaccination obviously provides a restraint on North–South travel, partly because it involves some physical discomfort, and partly because it involves a time lag for the first visit. This effectively rules out mass tourism. The problems of adjustment to the climatic difference for the traveler himself—although they may be serious in individual cases—are not generally serious enough either to warrant much concern. Indeed, in many cases the climatic difference may be the whole point of the trip, as in modem mass tourism from the Scandinavian countries to the Mediterranean in winter. In general, vertical distance is clearly a less important impediment to international interaction than horizontal distance. In the next section we turn to a closer examination of the problem created by time differences.

3.3 Time Difference and International Interaction

3.3.1 International Travel

Modern man increasingly lives by the clock, thus necessitating a stricter regulation of time. For an increasing number of people (although still a small minority in the rich countries and an even tinier minority on a global scale) *time* is replacing *money* as the most important scarce resource (Linder 1969). One of the strongest forces for standardization is precisely the improved means of communications. Each town or little area used to have its own time, but with the railroad this quickly became impractical. There were 75 different 'railway times' in the US before 1883, when US railroad managers set up their own standardized time zones for the purpose of simplifying their schedules. At the same time a movement for standardization was under way in Europe, motivated more by scientific than by commercial consideration. Eventually, within a few decades, most nations adopted one or more standard time zones (Schroeter 1926: II). A few countries, mainly Arab, still stick to 'sun time'.

A more basic standardization of time is, of course, the daily (diurnal, circadian) cycle. Figure 3.1 gives a generalized picture of the cycle. It applies to such bodily phenomena as *rectal temperature, heart rate, ion excretion*, as well as psychological phenomena such as *fatigue*. The periodicity is partly exogenous to the organism, regulated by such cues as light or darkness and human activities (eating, going to sleep), and partly endogenous, regulated by a biological clock with its own natural period.

A number of experimental studies on people living in dark caves without timepieces have confirmed that the natural period does not equal 24 h, which is why it is frequently called 'circadian'. One set of studies, for instance, determined the natural cycle to be 25.2 h (cf. Pöppel 1972). The endogenous signals can be modified experimentally and can adapt even to quite drastic changes. One such change is the modern East–West flight.

As any air traveler knows, rapid displacement over several time zones causes considerable discomfort. A number of recent studies indicate that at least part of this discomfort is associated with the time shift itself.[4,5] Estimates vary, but the evidence suggests that it takes anything from three days to a week to achieve complete readjustment after a transatlantic flight. For the return trip, readjustment is more rapid. The feeling of fatigue is overcome sooner than its physiological basis and some other psychological functions (decision-making ability, numerical ability) have not been shown to be decisively affected at all. The various studies are not unanimous in their assessment of the significance of time differences. Some

Fig. 3.1 Representative diurnal curve. *Source* Siegel et al. (1969: 6)

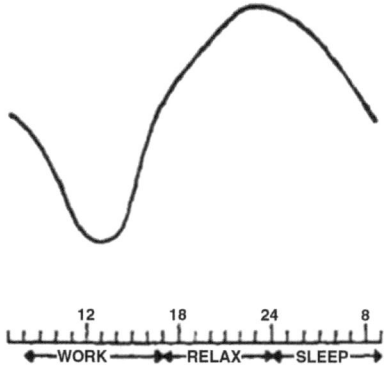

12   18   24   8

←—WORK—→◄►◄—RELAX—→◄—SLEEP—→

---

[4]Most of the research in this area has not been concerned with the well-being of the passenger, but with the fatigue of the airline crew. This is not an unimportant point in aviation safety: 'One BOAC pilot kept a careful log of his rest and sleep for 18 months … His passengers would not have been reassured to learn that in a representative spell of flying on North Atlantic routes this particular pilot had one period of sustained wakefulness lasting 23 h, another of 33 h except for a 2 h nap, and that he fell asleep for some minutes half an hour before the time for landing.'

[5]Blatt/Quinlan (1972: 507) distinguish between *dysrhythmia*—disparity between the internal clocks and the external temporal referents—and *desynchronization*—disparities between the internal rhythms. In this article we have not made this distinction.

researchers feel that loss of sleep and travel fatigue are more important than the circadian shift (cf. Evans 1970).

The readjustment problems vary with age, the most serious arising for persons whose daily cycle is closely regulated for medical or other reasons, e.g. diabetics.[6]

It is a disputed point whether East–West or West–East flights represent the greatest strain on the organism. On theoretical grounds, it has been argued that it should be easier to adapt to East–West flights because one can more readily suppress than advance sleep (and other periodic phenomena) for a few hours (Siegel et al. 1969: 7). However, the opposite theoretical prediction has also been made, and the experimental results are somewhat contradictory. A behavioral point which I have not found in the medical literature on this issue is that it is far more common to stretch out the day into the night (whether for business or social reasons) than to prolong the day by getting up very early in the morning. That this psychologically facilitates similar behavior after long distance flights seems likely. Furthermore, departures in the West–East direction are often in the evening, with arrival in the early morning. One reason for this is that a 6 h flight means a 12 h difference in local time (because one 'loses time' when travelling from West to East) and therefore difficult to fit a long flight into a normal day. West–East departures across the Atlantic, for instance, tend to be crowded into a few hours whereas the return flights are spread out over a greater part of the day. Psychologically, there is also a temptation to try to make up for the time 'lost' in West–East travel by travelling overnight.

The problem of rapid movement across several time zones is well-known to the experienced traveler. How large this group is, one cannot judge accurately. I am not aware of even rough estimates of what fraction of the population of the world or even of a particular country have had personal experience of the 'jet lag'.[7] However, the role of traveler is becoming sufficiently institutionalized for the

---

[6]The following informal rendering of a doctor's prescription for his patient's behavior following arrival in Rome at noon (local time) after a flight from Tulsa, Oklahoma may serve as an illustration: Upon arrival in Rome, sleep 2 or 3 h. Awake approximately 24 h after last daily injection of 22 units of NPH insulin. Run a sugar-urine test. Go downstairs and find out where a meal can be obtained so you will know where and when food will be available. Return to room, take 10 units of regular insulin. Eat a dinner within 30 min. Take a walk, see the fountains, sit down under sixteenth century arch to study the fourth century church, and listen to a twentieth century election campaign, loud PA system, records of choir singing. Before retiring, probably ten or eleven Rome time, run a sugar-urine test. Then run an acetone test. Ignore high urine-sugar under these circumstances, but if you show acetone, take 4 units of Regular-Insulin before going to bed. Keep sugar lumps on the table by your bed at night and in your pocket by day. Awake in the morning to a real Roman morning. Run sugar-urine and acetone tests. Take usual 22 units of NPH insulin. Eat breakfast Roman-style, enjoy the hard bread and the caffe latta (sic!) with your usual 2 units of protein. At this point, insulin time is synchronized with Rome time and you are on your own. *Source* Carney (1968: 10). Alternate plans deleted. This author accepts no medical responsibility for the plan!.

[7]The number of transatlantic passengers in 1971 was 11.3 million, rising 16 % p.a. over the preceding decade, according to IATA and related statistics. From the US we know that air travel has a very skewed distribution. In 1962 the top 25 % of business travelers accounted for 73 % of

problems to be felt as an institutional problem, too. Business organizations and foreign ministries and other bureaucracies with a high number of 'jet set' executives are beginning to have to face the problem. We shall return later to some of the solutions that have tentatively been introduced in order to deal with the problem. Let it suffice to note here that the diffused awareness of the problem makes it not unlikely that it may have a muffling effect on enthusiasm for international air travel across several time zones and that, conceivably, the effect may be read off directly in travel rates.

The International Civil Aviation Organization (ICAO) has summed up the strain on the international air traveler and the required rest period in the following formula, sometimes called Buley's formula (cf. Finkelstein 1972):

$10R = tt/2 + (tz - 4) + \text{dep coeff} + \text{arr coeff}$

(or $10R = tt/2 + \text{dep coeff} + \text{arr coeff}$ for trips across less than four time zones)

where

R is the rest period in days, rounded upwards to the nearest $1/\sim$ day

tt is travel time in hours,

tz is the number of time zones crossed,

dep coeff is a special departure coefficient

and arr coeff is a special arrival coefficient.

As an example, an air traveler leaving Montreal at 1800 local time is scheduled to arrive in Paris at 08:00 local time. The rest period is then $(9/2 + 1 + 3 + 4)/10 = 1.25$ or rounded off to 1.5 days. The two coefficients give some weight to departures and arrivals at inconvenient hours, to compensate for lost hours of sleep.[8]

The formula further gives greater weight to *travel time* generally, than to the East–West factor. Vibrations in the plane, the lack of movement in a restricted space, the drop in air pressure (even in pressurized cabins) and many other factors which apply to all flights, lead to a feeling of fatigue. However, it is at least conceivable that these factors may be eliminated by new technological developments. There is no similar way of eliminating the time difference although technological attacks on the effects of time differences are being attempted, too—as we shall see presently.

There is no built-in compensation for the effect of climate differences in the ICAO formula. A 1965 study carried out by the Office of Aviation Medicine of the US Federal Aviation Agency (Hauty/Adams 1965: 1) concluded that the North–South flight did not lead to a shift in the circadian cycle, but that it did lead to an

---

(Footnote 7 continued)

all business air trips and the top 6 % for 26 % of all non-business air trips (Lansing et al. 1964: 96). However, one can only guess how many *different* people have crossed the Atlantic in a given year.

[8]Thus, a poor arrival time on any other standard can be made into a virtue. An SAS advertisement explains that arriving in Tokyo 09:05 on Sunday morning is just ideal since it gives one a whole day to relax and adjust.

Table 3.2 Seven measures of distance and one form of international interaction, variable definitions and source

Concept	Measure	Year	Source
Horizontal distance	Difference in longitude	1969	Gleditsch (1969)
Vertical distance	Difference in latitude	1969	Gleditsch (1969)
Bee-line distance	Great circle distance	1969	Gleditsch (1969)
Climate difference	Similarity/non-similarity on Köppen's climate scale	1960	Köppen (1900); Rumney (1968)
Wealth difference	Difference in GNP/cap	1965	UN statistics
Travel fatigue	ICAO formula		See above, p. 43
Time difference	Absolute difference between time zones	1969	Time zones coded from airline schedules and reference works
International flights	No. of weekly scheduled flights between the two countries		Gleditsch (1969)

increase in 'subjective fatigue'. In no study that I have seen have the three effects of time difference, climate difference and travel time been systematically untangled.

There is, of course, nothing magical about the ICAO formula. However, it sums up the perception of an important international agency of the magnitude of the problem.[9] Conceivably, there could be important effects from the perception in the travelling community of the effects of time differences, even if the image did not have sound medical bases. The only way to study this would be by a thorough examination of travel rates in all directions. Here, we shall only report a preliminary test.

We have computed correlations for seven measures of distance (functional and otherwise) for one form of interaction, international flights. Table 3.2 surveys the variables.

Correlations were computed between these variables for 24,090 nation dyads and a typal analysis performed on the correlation matrix. Figure 3.2 depicts the statistical relationship between the variables. The only high correlations (>0.75) are between great circle distance, ICAO travel fatigue, East–West distance, and time difference. The correlation between East–West distance and great circle distance

---

[9]However, the acceptance is not unanimous. Secretariat members tend to push for longer rest periods, whereas the personnel office tends to prefer more conservative estimates. Other special agencies of the UN have not, so far, accepted the ICAO formula, nor does this appear to be the case in national bureaucracies with much international travel. The importance of rest is well known, of course. In travelling to China in 1972, President Nixon made overnight stops both in Honolulu and Guam.

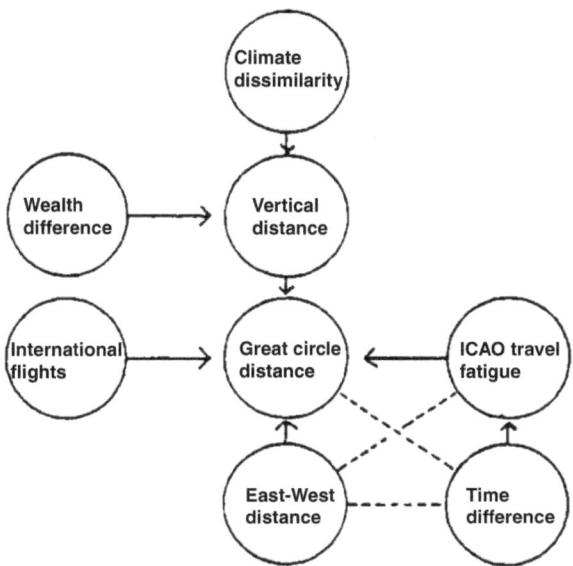

Fig. 3.2 Statistical relationship between seven measures of distance and international flights, 1965. Based on product-moment correlations and a 'typal analysis' (McQuitty 1961) of the correlation matrix. The relationship A → B means that variable A has a higher correlation with B than with any other variable (negative correlation in the case of flights). Broken lines indicate other correlations ≥0.75. All remaining correlations are ≤35. The full correlation matrix is reproduced in the Appendix

(0.9) greatly exceeds that between North–South distance and great circle distance (0.35). This, incidentally, tells one something about the geographical structure of the world: most countries are distributed along a relatively broad band on both sides of the Equator. There are no capitals north or south of the polar circles and very few north of 60° or south of 40°. Hence, if travel was randomly distributed between countries, there would be more travel in the East–West direction than North–South. In this sense also, time difference becomes a more important problem in international interaction than climate difference.

As one would expect, *climate dissimilarity* and *wealth dissimilarity* are statistically associated with *vertical distance*. However, the linear correlations are generally low.

The correlations with the indicator of international interaction (international flights) are not high either. In a linear model, the distance measures have a multiple correlation of only 0.2 with flights. Previous studies have shown that multiplicative models of size and distance variables usually account for more variance in interaction than linear models (Gleditsch 1969 and references therein). However, multiplicative models have not been tried out in this case. Great circle distance and East–West distance are the two variables which together account for most of the variance in flights. But other variables beyond the first add little to the variance reduction.

As is evident from a glance at a world map, more countries extend in the North–South direction than East–West. Hence, for purposes of intra-national travel, vertical distance is more relevant than horizontal. (Where the 'social axis' of the country is East–West, even though the country extends geographically North–South —such as in Chile—the East–West distances are likely to be small, and no problem arises.) But more important and for precisely the same reason, because the three great oceans (Indian, Atlantic, and Pacific) divide vertically between nations rather than horizontally, distances between nations tend to be horizontal rather than vertical. It is to be expected, then, that great circle distance and East–West distance should be highly correlated (0.9). However, the lack of data on national product for many countries in the South and the lack of a climatic classification for a number of countries, exaggerates this correlation somewhat. Countries for which no data were available on one or more variables had to be excluded from the analysis, and more of these were in the South than in the North.

This tentative analysis supports the greater emphasis of the ICAO formula on geographical distance than on East–West distance. However, in a stepwise regression analysis, East–West distance was found to be the second most important predictor of flights. We conclude then, that while there are certainly many other important factors in determining global interaction patterns, medical research and a preliminary analysis of interaction rates agree that geographical distance is still an impediment to interaction and that horizontal distance is a more important impediment than vertical distance.

The increasing *speed* and decreasing (relative) *cost* of travel, have removed some of the negative effects of distance but the problem of travel fatigue, etc., in long-distance travel generally and East–West travel in particular, is only aggravated by the same trend.

One possible solution to the whole problem would be to reduce the importance of travel, and rely more on telecommunication, communicating symbolic information rather than persons. This possibility is examined in the next section.

3.3.2 Communication without Travel

A great deal of technological innovation is now geared toward this end. The letter and the cable were the first primitive steps and both are now rapidly decreasing in importance. The telephone and telex are currently the most important modes of communication for decisions of great significance or decisions which have to be made fast. Telex makes possible a practically instantaneous communication of written messages. The telephone message permits greater flexibility because it is oral and the lack of any record of the message permits greater freedom of expression. However, telephone messages can also be recorded for future use if desired (and, as we have now learned the hard way, even if not desired). The conference phone extends the use of this medium from bilateral to multilateral conversations. The conference videophone is merely an extrapolation of technological capabilities,[10] while the smelling picture phone probably requires some new technology (and, in any case, seems less important). Communication satellites and other technological innovations have multiplied the number of available channels for long distance telecommunication. Telecommunication networks no longer know any insurmountable hurdles in terms of geographical distance, although the quality of the network varies considerably with, e.g., the economic level of different countries. In effect, then, quite complex transactions can be carried out without moving the agents physically.

However, there are limitations to this. As any businessman knows, difficult negotiations can be helped along by a good dinner, a relaxing drink or a lively evening out. Electronic communications media have no satisfactory substitute for these icebreakers. To a certain extent, just as the telephone and telex have replaced the letter as the business mode of non-personal communication, the airplane has replaced the railroad and the ship in the movement of persons. This does not necessarily alter the mix of face-to-face and non-personal communication. Technology has an edge now, since a phone call or a telex message can be relayed faster than one can fly to the point in question. In the old days a letter was no quicker than the steamship or the railway and then one might as well have travelled in person. Furthermore, the telephone (but not the telex) will permit instant two-way communication, which comes much closer to face-to-face communication than the letter does.

But in actual fact, the two modes of communication appear to increase in volume together, rather than compete for demand. Routine business and items which require a very fast reaction can be handled by symbolic communication, while more basic transactions which take time anyway can still best be taken care of by face-to-face communication. The initial 'acquaintance process' in particular needs face-to-face communication. If at least one of the interacting partners belongs to a culture with a high personnel turnover in the relevant roles, the need for face-to-face meetings in order to 'get to know each other', will be more or less constant. Besides, vital messages cannot always be sent by telephone if the line is not secure.[11]

---

[10]A trial effort has been set up by the General Post Office in Britain under the name of 'Confravision'. So far the system only links Gresham St in the City of London with a post office research station near Wembley. However, a fully developed UK network is being planned, complete with tape recorders, photocopying machines, and scramblers. The suggested price is 120 pounds per hour between London and Manchester. An interesting psychological point in connection with 'Confravision' is that it will—at least in the beginning—impart a sense of urgency and contribute to a streamlining of the discussion (cf. Baxter et al. 1970: 93–97).

[11]'Office work is conducted with only the rarest recourse to the telephone. Washington does not call because when it is noon in Washington it is midnight in New Delhi, give or take an hour or so, and the line is not secure' (Galbraith 1970).

Thus, we do not believe that the problem of overcoming the restraining effect of distance in general and time differences in particular can be overcome by switching to non-personal communication. However, it is possible that the ratio of non-personal to face-to-face communication may increase somewhat.

This does not eliminate the problem of time difference, however. Time differences also mean non-overlapping working hours. There is no overlap in the normal working day between Tokyo and New York City or between Tokyo and the capitals of Western Europe.[12] This means that one cannot use the telephone for less than vital messages. A telex message from London to Tokyo will arrive after hours and can only be dealt with the next day. When Tokyo gets around to replying, London is off work. The effect of all this is that we are back to one-way communication patterns which are not significantly faster than the movement of persons. Thus, if the matter is urgent, an executive may as well fly from New York to Tokyo or vice versa for a consultation, rather than send a telex message.[13]

The exception to this rule lies, of course, in the possibility of calling outside normal office hours. There are strong norms against doing this within the same city or country where there is no time difference. There is a strong and mutual interest among decision-makers, business or otherwise, in safeguarding the privacy of one's home (or the privacy of one's spare time). In international communication, however, the problem of non-overlapping business hours complicates the issue.

As technological possibilities improve, two-way communication (phone calls) outside regular business hours will no doubt become a more common phenomenon. At the same time, the volume of one-way communication will also increase. The wire services will send out messages on a 24 h schedule, particularly as events that are universally defined as 'news' are produced in more countries all over the world. There will be increasing pressure on the radio and television stations in all countries to continue to extend their program time (as they have done in the past) for news programs, but also for relaxation, as the number of working hours decreases. There may be political decisions to delay events by a suitable number of hours to fit the news schedules in the receiving country, as the introductory example of the Olympic Games demonstrates, but as one medium competes with another the situation will be increasingly difficult for those who always lag behind. In the case of the Norwegian coverage of the 1972 winter Olympics, for instance, the main results would frequently be known from the radio before the television program even started. Enthusiasts who listened to East German radio might even have heard the results before going to bed at 1 am. In the long run, this kind of competition will probably give a country or a medium the image of always being 'last with the news'

---

[12]Among the 219 nations and territories in the world in 1969, the number of non-overlapping countries ranged from 30 (France and other countries on Central European Time) to 169 (Fiji and others). This was calculated with exact knowledge of the time position of each country but with the naive assumption that all countries had working hours from 09:00 to 17:00 local time.

[13]The working day is de-synchronized even more by the variation in lunch hour habits. In Europe alone, the lunch break varies in length from 20 min to 3½ h, and nominal starting time is different from country to country, too.

and it is difficult to imagine that this will not lead to pressure—internal as well as external—to take up the challenge.

Business information will be a particularly important form of news transmitted on a 24 h basis. The Associated Press/Dow Jones Economic Report, for instance, provides around-the-clock information from news rooms in New York and London. Once the information is available in a country (as it was in Norway in 1971–72) there may probably be local pressures for dispersing it within the country.

Clearly, the problems associated with time differences call for inventiveness, technological or social or both. But before we attempt a systematic discussion of possible (desirable and undesirable) responses, we shall give two examples of how two large organizations have experienced these problems and attempted to handle them.[14]

3.4 Examples

3.4.1 An International Business Organization

The domination of one wall in the telex room by half a dozen clocks indicating the time in different places all over the globe, serves as an instant reminder that a shipping organization is extremely sensitive to the problem of time differences. Executives, in fact, call the telex room quite frequently, asking the time in Tokyo or Cape Town, rather than taking the risk of making an error in their private calculations. The shipping market is an international one and its centers have shifted after the war from Europe (particularly London) to New York and Tokyo. Moreover, it is a market where big decisions have to be made fast.[15] This is particularly true in cargo chartering (as opposed to the line trade). A customer may want to charter a 100,000 ton ship for 3–5 years and the decision has to be made while the agent is on

---

[14]In addition to non-overlapping working hours, there is also the problem of non-overlapping public holidays. For instance, Norway has three public holidays at Easter—not counting Palm Sunday and Easter Sunday—while many other Christian countries have none. The New Year is celebrated at several different times in different cultures. Goody (1968: 37) reports that Brazil has 18 bank holidays, Britain 6, and Bulgaria 5. The example indicates a certain rationalization in 'modern' societies (i.e. in this context, societies where time is scarce). Throughout antiquity and the Middle Ages, there used to be no less than 115 public holidays during the year (Craven 1933 [quoted from Linder 1969: 26]). Now, public holidays are removed while vacations are increased. Still, a major financial institution has found it necessary to issue a list of Bank and public holidays throughout the world (New York, Morgan Guarantee Trust Company, Ltd. 1965). And in Norway in early 1972 an international gang systematically exploited a Swedish bank holiday to pass forged Swedish checks in Oslo banks (cf. *Aftenposten*, ev. ed. 7 February 1972).

[15]This is not limited to shipping, of course: 'In 2 min this man buys and sells more money than you could make in twenty lifetimes. The man is Jan Gorski. Chemical Bank's chief foreign exchange trader in New York. In 2 min, recently, Jan and one of his staff bought and sold one hundred million Deutsche marks ...' (from a Chemical Bank advertisement).

the phone from New York or Tokyo. The price of the phone call is not a factor in cases like these, but time is.

In order to cope with increasing international communication, this shipping agency has joined with a Japanese firm in an interesting international operation. The working day for this international firm starts in Tokyo. At the end of the working day in Tokyo a long report is sent by telex to Oslo. Here, the information is passed directly on to London (technically this is done by feeding the output paper tape from the telex into the input tape reading unit of the telex—thus, only a minimal time lag is involved). At the end of the Oslo working day information and control are passed on to New York and then again to Tokyo for a new day. This is, in a sense, a form of international three-shift working system. However, the three centers are not completely equal. Oslo wants to retain ultimate control and Tokyo and New York (and London) only have limited discretion. For bigger contracts, the branch offices have to consult with their counterparts in Oslo. The instant decision on a 100,000 ton ship may therefore have to be made at two o'clock in the morning. Not only does the shipping agency maintain a regular telex watch until seven o'clock in the evening, but quite frequently the telex lines well be held open far into the night if an important message is expected. Executives call the cable office at seven o'clock and at ten o'clock in the evening (as well as three times on Saturdays and twice on Sundays) to check incoming cables. The cable office also holds standing instructions to call executives at their homes if important cables are received (such as notification of accidents).

An interesting aspect of this particular international operation is that an organization in a small and peripheral country is the dominant center of the joint operation. The normal pattern would be for the dominant partner to be located in a dominant country (e.g. the IBM or the ITT with their head offices in the US and a number of branches in foreign countries). In this case, there can be little question as to *who wakes whom*: The dominant partner contacts the dominated partner at an hour convenient to the former, but there is no disturbing the peace in reverse unless absolutely necessary. In other words, the dominated and peripheral partner yields to the daily cycle of the dominant and central partner. However, in the case just mentioned, the dominant partner has to adapt to the time pattern of the dominated. The country dominance factor appears to be more important than the organizational dominances. In part this may be because organizations in dominant countries are used to determine other people's time, rather than have the reverse happen to them. But a more important explanation is probably that the organization in the peripheral country has to adjust to the market, which in turn is adjusted to the dominant country. Of course, this may vary between one form of interaction and another. In shipping, the central decision-maker must be available at all times. In other multinational ventures, it may be more important for the peripheral parts of the organization to be available to respond to the whims of the center.

3.4.2 The Foreign Ministry of a Small Country

The structure of information gathering in foreign ministries has changed in the postwar period with decreasing emphasis on the traditional formal reports and increased use of telex, cables, and telephone communication. Personal meetings—often in stopovers before or after another meeting or conference—are also more frequent and, as a result, much less formal than before (fewer dinners, more political talks).

However, so far this has not led to any drastic change in the working habits of the Norwegian foreign ministry. Evening meetings are quite common—two to three times a week at the top level in the fall season (which is the busiest because of the UN General Assembly). But the overwhelming number of such meetings are held because of a general scarcity of time and because other commitments sometimes do not permit undisturbed meetings during the office day.

Even so, it is not unusual for foreign ministry officials and politicians to have to make sudden decisions outside working hours, and even in the middle of the night. This is particularly frequent during sessions of the General Assembly. But rather than channeling new instructions through the regular foreign ministry network, a member of the New York delegations will call a top official or politician at his home directly.

So far only the UN has presented a significant number of problems which demand an urgent decision. But similar problems were anticipated at the time of the interview in connection with the UNCTAD III conference in Chile, '9 h away' from Norway.

Bilateral contacts with other foreign ministries by telex, telephone or cable are extremely rare. Personal meetings on a bilateral basis are somewhat more frequent, but less numerous than multilateral contacts (in the UN and NATO particularly, now also MBFR, CSCE, etc.). But the traditional channel of communication, through the embassies, still remains the most important.

Intra-Scandinavian contacts are an exception. Here, telephone calls are quite common, from the foreign ministry level down to preparatory clerical level. Ordinary telex lines are also used frequently. There is a great deal of mutual confidence, and officials are often on first name terms. When such personal contacts are made outside Scandinavia, on the other hand, there is something dramatic about them, they are news. A prime example was a cable sent in December 1971 to the Norwegian Prime Minister from the British Prime Minister urging moderation in the negotiations with the European Community.

Compared with international business decision-making, as described above, international foreign policy decision-making takes on a somewhat old-fashioned tinge. One cannot escape the feeling that just as the intra-Scandinavian embassies are increasingly irrelevant for important political decisions, extra-Scandinavian contacts will also have to be made on a more direct basis at some point. This is, of course, particularly true for crisis situations but even in non-crisis times the contrast with business decision-making and with intra-Scandinavian decision-making will probably make itself felt.

Another structural factor of some importance is that the foreign ministry is centralizing. The central administration is strengthened and the embassies to foreign countries—as distinct from representations at international organizations—are reduced in importance, if not in number. (In any case, their number has not kept pace with the growth in independent countries.) At the same time, increasing foreign travel for diplomats stationed at home is explicitly foreseen. This means that the problem of long-distance travel will have to be faced more seriously. So far, no formula like that of the ICAO has been applied by the foreign ministry and an official who wants to extend a visit in order to incorporate a rest period, may run into administrative problems. At the same time the lack of permanent representation in a number of important cities will necessitate more frequent use of telecommunication.

3.5 Technological 'Solutions'

If a worn-out phrase like 'a technological age' should be applied to our time, the most appropriate reason would seem to be our tendency to look first for technological solutions whenever new problems arise. This is particularly true when the problem occurs in the first place as the result of a technological innovation. Even though, as we have stressed before, the problem of time differences is only beginning to make itself felt, a number of wheels have already been set in motion to break the back of the problem.

As the previous discussion indicates, there are two separate problems: (1) the circadian shift in international travel, and (2) the problem of non-overlapping working hours in telecommunication. Of the four technological solutions we shall discuss in the following, the first three refer to international travel, the last to telecommunication.

3.5.1 The Pill

If there were no diurnal cycle, there would be no problem. But man would also be a rather different animal. I have not come across any suggestion to eliminate the diurnal cycle. But travelers are frequently advised to try to adjust to the problems of time shifts (don't drink too much, don't eat heavy meals, depart in a rested state, rest after arrival, take it easy during the asynchronous period, etc.). Apparently all this is not enough, for there is at least one project to develop 'a pill' which will expedite the adjustment of the cycle to a new time zone. No serious discussion of the effectiveness of such a pill or of possible side effects has yet come to my attention but as a skeptical layman I feel inclined to think that there must be side effects which would probably be harmless for a few trips per year across the Atlantic, but not for travel several times per week.[16]

3.5.2 Transcendental Meditation

Recently, according to a report in the *New Scientist*, a number of papers have appeared in learned journals about the physiological changes accompanying transcendental meditation. These changes include a decrease in metabolic rate and in breathing rate, an increase in skin resistance, a reduction in heart output, etc. In short, in the words of a leading spokesman, 'the body is deeply rested, but the mind remains alert'. TM has been used to cut down on smoking and the use of tranquilizers and stimulants and even drugs. The latest claim is that it enables its practitioner to avoid the unpleasant consequences of jet lag.[17]

3.5.3 Sticking to One's 'Home Time'

For those who frequently travel across several time zones, the best strategy may be to try to stick to their 'home time' and make their stay short enough to return to base before the organism has time to adjust.[18] This in a sense is a social and not a technological solution, but it is one which is backed up with a great deal of technology and is therefore dealt with in this section. The most important technological innovation geared to this strategy is probably the SST. An objection often made of the SST project is that it does not make much difference if one crosses the Atlantic in 3 h rather than seven. The objection would probably be muted if it was a matter of cutting in half one's own daily travel time to work or even the weekly travel time to a summer place in the country, etc. Furthermore, many flights are considerably longer than the Atlantic crossing. By late 1972, the Los Angeles to Hong Kong record (7,677 miles) was 14¾ h. But what is perhaps more important, the SST opens up the possibility of travel, e.g., across the Atlantic on one's home time. Table 3.3 sets out a hypothetical travel and conference schedule from New York to

---

[16]The development of such a pill at the Syntex Corporation appears to be running into difficulties, and marketing of such a product is way into the future. Kahn/Wiener (1967) also mention 'controlled or supereffective relaxation and sleep' as likely inventions in this century.

[17]New Scientist (1973).

[18]Henry Kissinger, in twelve secret visits to Paris during the Vietnam negotiations, made the trips so short that his absences from Washington would not be noticed (and were not!) He 'kept his watch on Washington time' in order to minimize the effects of the time lag. A few of these round trips were completed in 22 h, and he occasionally arrived back in Washington so late that the 'post-mortem' with President Nixon was held in the latter's bedroom. *Time Magazine*, 7 February 1972.

Table 3.3 A hypothetical day trip to London for a busy American after the introduction of SST	New York local time	Scheduled item	London local time
	8	Depart	13
	11	Arrive	16
	15	Depart	20
	18	Arrive	23

London and return. A busy American can travel to London, have a three hour conference, and return to New York, all in a not unreasonably long day.[19]

It can be argued that this type of lifestyle applies only to a minute fraction of the population in the two countries involved and that the benefits are greatly outweighed by the environmental hazards of the SST, which have consequences for nearly everyone. This would be a valid point if there were to be a referendum on the SST, but it carries relatively little weight if all the relevant decision-makers belong to the jet set.[20]

With the reduction of flying time, ground time occupies an increasing fraction of total travel time. There will also be consumer pressure for simplification of airport procedures and for more efficient mass transit between airports and population centers. The increasing size of airports and the tendency to locate them further away from cities will, however, work in the opposite direction.

The 'sticking to one's home time' strategy is being applied already, even without SSTs. According to a news report,[21] the International Telephone and Telegraph Corporation holds monthly European meetings for 150 top managers from Europe and the US.' The windows are curtained to banish time. Most members of the Manhattan contingent, who fly over by chartered Pan American 707 jet, keep their watches on Eastern Standard Time.[22] There is no smoking allowed and only mineral water is available.

---

[19]Although this example was made up, I was gratified to find later that the Director-General of SAS has made the same point: 'What is really the advantage of SSTs?—You fly to America in about 4 h and can get back the same day after lunch and work. This is what you do today when you go to London, Stockholm, Copenhagen, or Paris. I think it will be just as necessary and natural in 10 years to fly to the US on the same basis.' (Hagrup 1973).

[20]For those who argue that the high costs of supersonic service will prove its demise, a US survey by Market Facts (as reported in *Flight International*, 10 May 1973) offers scant hope: 239 business travelers were interviewed. They had made a total of 743 business trips in the previous 12 months, of which 553 were economy class. Some 70 % of the economy class passengers indicated that they were prepared to pay a 40 % higher fare to fly in a one-class Concorde. It was further estimated that 12 % more trips would have been made in 1972 if a Concorde service had been available. In other words, the business community is highly responsive to cuts in *travel time*. The explosion in charter traffic shows that there is another market which is more responsive to *price cuts*.

[21]Story and quotation from *Time Magazine*, 20 December 1971.

[22]Provided, of course, that they have not purchased the new Accutron wrist-watch with 2 h hands, one of which 'tells the time where your mind is', the other 'the time where your body is' (as

This strategy is only possible if the stay is made relatively short and even then it helps to shut out external cues like daylight, which would set the cycle change in motion. For industrial workers on all-day shifts, such a strategy may actually have negative effects: 'Industrial workers who start shift work often change their sleeping hours only on days when this is imperative and on off-days continue to sleep at night, trying to live according to normal time routines whenever possible, with the result that they never adjust. Consequently, their circadian temperature may merely flatten, leaving them below their best potential while at work.'[23]

Finally, while 'solving' the problem of adjusting to international travel, this strategy shifts the problem into the other area discussed previously, that of non-overlapping working hours. If the traveler will not yield, his host may have to.[24]

3.5.4 Changing the Light Cycle

It would be simple, of course, if one could change the cycle of night and day so that all countries would be on exactly the same schedule. In theory there is no great problem in designing a set of gigantic mirrors which would reflect the sun evenly all over the globe at the same time (and perhaps absorb the energy from the light for the night period). In practice, this idea belongs to science fiction, although the idea of illuminating parts of a country was discussed in connection with the Vietnam War.

3.6 Social 'Solutions'

Technological solutions to the problem of time differences appear to raise (at least) as many problems as they solve. But are there any social responses, solutions that involve a particular organization of world society or a particular life-style for its members? Of course technological innovations may be part of such solutions, but the focus here is on the social innovation. All of these 'solutions' apply to the

---

(Footnote 22 continued)

advertised in *Playboy*). Or the $575 computerized Bulova clock which on demand will flash the time of day in any major capital (as marketed by a US airline) etc.

[23]The quotation is from an editorial in the *British Medical Journal* (1970: 760). The three-shift system, 1 week to a shift, may be the worst of all possible systems. If it takes a week to adjust to an 8 h change, the shift worker will be in a constant state of adjustment to his current schedule.

[24]'Sticking to one's home time' as a response to time differences has a parallel as far as climate differences are concerned: the increasing use of climate control for buildings and even whole cities, will enable the traveler to stay in his 'home climate' or at least a 'standard international climate' for the duration of the trip.

problem of non-overlap in working hours. The third and the fourth are also relevant to the problem of travel.

3.6.1 Time Imperialism[25]

The most likely development seems to be *the imposition on small countries of the time cycle of the dominant countries*. It is illuminating that for branches of international businesses located in Norway—whether dominated or dominant inside the organization business—*the Norwegians are the ones to be awakened*. International stratification seems to take precedence over organizational dominance. The economic and political dominance system of the world extends itself into other areas like 'time'. It is typical that when the standardization of time on a world-wide basis was first seriously discussed in the 1880s, one of the more prominent proposals was for a *world day* which would begin with midnight in Greenwich. The proposal was rejected, however, because of the inconvenience of beginning work at nominally different hours in different places. The actual outcome of the standardization—the 24 time zones—is also centered around the observatory in the capital of the dominating country of the day—but in a less drastic way. Time was not ripe for the more drastic solution of the 'world day'.[26] In a few years' time, however, the world may be ripe (not by decision, but through practice) for an infinitely more drastic

---

[25]The use of the word 'imperialism' in this context should be understood as a characterization of inter-personal relations as much as inter-nation relations. Raimo Väyrynen has criticized the present paper for directing attention to the problems which still mainly concern a tiny élite whereas the problems of three shift workers, etc., are discussed only parenthetically. If time imperialism were the central focus of this paper, this criticism would be justified. In 1971 it was found in a survey of three Norwegian male cohorts (1921, 1931, 1941) that 10–15 % of those employed had irregular working hours, with little variation between cohorts. Nearly 10 % were weekly commuters. (Unpublished data from the Norwegian Occupational Life History Study, Institute of Applied Social Research.) It must be assumed that for the vast majority of them this irregularity is other-imposed rather than self-imposed, as for many artists, intellectuals, etc. Whether domestic time imperialism will still be more important than the inter-nation variety in the long run (e.g. half a century from now) I feel less certain about. At least the two will become more strongly intertwined. It should also be remembered that even if long-distance travel and occupations with international involvement are (still) mainly concerns of the elite, international mass communication is not. The TV and radio transmission of the Olympic Games were mentioned initially. Other programs that come to mind are 'European pop jury' and the Eurovision song contest. And in the future the global song contest and international televised political debates? If the Security Council and the General Assembly of the UN were more like national parliament in that they made decisions that were (a) important, (b) in some degree unpredictable, then there would be a better case for televising such debates globally.

[26]The argument which could still be advanced in favor of such a proposal is that it might serve to increase global awareness and belongingness.

solution where people's daily cycle of work, sleep, etc., is not governed by light and dark but by the work habits of their counterparts in a dominant country.

It is hard to imagine that a major part of, say, the Norwegian population increasingly will work at night. A more likely outcome is that the distinction between leisure time and work time will be eroded for more and more people in the dominated countries—particularly for those in the service sectors which cater to an international market and in industries dominated by foreign interests.

3.6.2 Multilateral Control over Decision-Making

If control over decisions could be shared multilaterally, there might not be need for more than periodic consultation between the various centers. The example of the shipping organization suggests such a solution: when normal working hours close, all relevant information is transmitted to a point where the working-day is starting and from then decisions are made there until it's time to pass everything on to the next point, etc. The example discussed also suggests the problem with this procedure, however: the dominant center does not trust the other points to make the right decisions. They are therefore granted limited authority. Solving this problem is about equivalent to any other problem of real equality in organizations, no more, no less.

3.6.3 Move Everyone (Who Counts) to the Same Place

A curious aspect of the modern nation-state is the concurrent development of modern communications technology (which in theory should permit rapid and effective communication between all parts of the country) and concentration of population in a few great centers. If the same development occurred at the international level, it would certainly solve the problem of time differences. It is not that everybody has to move, only the people who want to be party to the decisions that are made at the international level.

3.6.4 Ignore Time Differences

While the problem of time differences still only affects a small part of the population, there is no necessity that it will spread to larger groups. A counter-culture may arise deciding to fight the kind of lifestyle which creates the problems discussed here. Such a counter-culture might prefer a certain reduction of material welfare if this meant less of a scarcity of time, a less hectic life, etc. The answer to the problem of time differences in this kind of culture would be: *so what?* A letter would be considered fast enough for non-personal communication and a ship would be fast enough for travel. The perspective would then not be that of solving the problem of time differences, but that of actively fighting the life-style of the jet set.

Appendix: Correlation Matrix for Seven Measures of Distance and One Form of International Interaction

		1	2	3	4	5	6	7	8
1.	North–South distance	1.00	–	–	–	–	–	–	–
2.	East–West distance	0.08	1.00	–	–	–	–	–	–
3.	Great circle distance	0.35	0.90	1.00	–	–	–	–	–
4.	Climate similarity	–	−0.16	−0.07	−0.10	1.00	–	–	–
5.	Wealth difference	0.32	−0.01	−0.00	−0.12	1.00	–	–	–
6.	Travel fatigue	0.24	0.88	0.95	−0.09	−0.00	1.00	–	–
7.	Time difference	–	0.11	0.78	0.80	−0.07	−0.01	0.01	1.00
8.	Flights	−0.11	−0.11	−0.16	0.08	0.03	−0.13	−0.12	1.00

(n = 5,671)

References

Aschoff, Jürgen (Ed.), 1965: *Circadian Clocks* (Amsterdam: North Holland).

Baxter, Raymond; Burke, James; Latham, Michael (Eds.), 1970: *Tomorrow's World* (London: British Broadcasting Corporation).

Blatt, Sidney A.; Quinland, Donald M., 1972: "The Psychological Effects of Rapid Shifts in Temporal Referents", in: Fraser, Julius T.; Haber, Francis C.; Müller, Gert H. (Eds.): *The Study of Time* (Berlin: Springer): 506–22.

Carney, Helen K., 1968: "Insulin Time Versus Greenwich Time", in: *American Association of Industrial Nurses Journal*, 16,8: 10–11.

Central Bureau of Statistics, 1972: *Seerundersøkelse om Fjernsynsteatret februar/mars 1973* [Viewer Report on the TV Theatre in February/March 1973]. Report (18) (Oslo: Statistics Norway).

Craven, Ida, 1933: "Leisure", in: *Encyclopedia of the Social Sciences* (New York: Macmillan): 402–05.

Curlis, C.J., 1949: *The Day of the Two Moons* (Washington: Association of American Railroads).

Evans, J.I., 1970: "The Effect on Sleep of Travel Across Time Zones", in: *Clinical Trials Journal*, 7,5: 64–75.

Finkelstein, Silvio, 1972: "Rest Can Help Travellers After Long Flights", in: *ICAO Bulletin*, 27,11: 10–11.

Fraser, Julius T.; Haber, Francis C.; Müller, Gert H. (Eds.), 1972: *The Study of Time* (Berlin: Springer).

Galbraith, John K., 1970: *Ambassador's Journal* (New York: Signet).

Gleditsch, Nils Petter, 1969: "The International Airline Network: A Test of the Zipf and Stouffer Hypotheses", in: *Papers, Peace Research Society (International)*, 11: 123–53.

Gleditsch, Nils Petter, 1967: "Trends in World Airline Patterns", in: *Journal of Peace Research*, 4,4: 366–408.

Goody, Jack, 1968: "Time", in: *International Encyclopedia of the Social Sciences* (New York: Macmillan).

Hagrup, Knut (interview), 1973: "Flyindustrien Lever Marginalt [The Aviation Industry Lives on the Margins]", in: *Norges Industri*, 55,5: 18–24.

Hauty, G.T.; Adams, T., 1965: *Phase Shifts in the Human Circadian System and Performance Deficit During the Periods of Transition: I. East-West Flight* (Fort Belvoir: Department of Transportation, Federal Aviation Agency, Office of Aviation).

Kahn, Herman; Wiener, Anthony, 1967: *The Year 2000* (New York: Macmillan).

Köppen, Wladimir, 1900: "Versuch einer Klassifikation der Klimate, vorzugsweise nach ihren Beziehungen zur Pflanzenwelt [An Attempt to Classify Climates, Primarily with Respect to Their Relation to Plants]", *Geographische Zeitschrift*, 6,11: 593–611.

Lauschner, Erwin A., 1971: "Luft- und Raumfahrtmedizin—Flugmedizin [Aeronautic and Astronautic Medicine—Aviation Medicine]", *Gesundheitspolitik*, 13,506: 259–79.

Latham, Michael (Ed.), 1970: *Tomorrow's World* (London: BBC).

Linder, Staffan B., 1969: *Den rastlösa velfärdsmänniskan. Tidsbrist i överflöd—en ekonomisk studie* [The Harried Leisure Class] (Stockholm: Bonniers).

McQuitty, Louis L., 1961: "Elementary Factor Analysis", in: *Psychological Reports*, 9: 71–78.

New Scientist, 1973: "Meditation Can Be Scientific", in: *New Scientist*, 58,847: 501.

Pöppel, Ernst, 1972: "Jet Travel—Body and Soul", in: *New Scientist*, 55,807: 232–35.

Rumney, George R., 1968: *Climatology and the World's Climates* (New York, London: Macmillan).

(UNRISD), 1969: *Compilation of Development Indicators for 1960* (Geneva: United Nations Research Institute for Social Development).

Schroeter, Jens F., 1929: *Haandbog i kronologi* [Handbook of Chronology] I–II (Oslo: Cammermeyer).

Siegel, Peter V.; Gerathewohl, Siegfried J.; Mohler, Stanley R., 1969: *Time Zone Effects on the Long Distance Air Traveler. Department of Transportation* (Washington, DC: Federal Aviation Administration, Office of Aviation Medicine).

Stoltenberg, Thorvald, 1970: "Avgjørelsesprosessen og Utenriksdepartementets plassering [The Decision-Making Process and the Role of the Foreign Ministry]", in: Stokke, Arne J. (Ed.): *Beslutningsprosessen i norsk offentlig administrasjon* (Oslo: Universitetsforlaget): 106–122.

Chapter 4
Democracy and Peace

The observation that democracies rarely if ever fight each other was made by Dean Babst nearly three decades ago, but has had little impact on the literature on peace research and international relations until recently. But now every volume of the leading journals contains articles on minor and major aspects of this theme. Professional jealousy and confusion of levels of analysis are possible explanations for the late acceptance of the idea of a democratic peace, but above all it seems to have been hampered by the Cold War. Erich Weede has taken a bold step in reconsidering his own previous view and other should follow. The Cold War has ended in the real world, and it should end in peace research, too.

'Democracy encourages peaceful interaction among states'.[1] This proposition flourished during the Enlightenment—it was, for instance, a central part of the political debates which surrounded the American and French revolutions. As early as 1795 Immanuel Kant described a 'pacific federation' or 'pacific union' created by liberal republics.[2] Much more recently this topic has become the subject of systematic empirical observation. In this issue of *JPR* we publish four articles on the relationship between democracy and peace. In the first, Weede (1992) reconsiders his previously published view (1984, 1989) that extended deterrence and subordination to superpowers are the major pacifying conditions in the international system. He now joins the emerging consensus that 'democracies do not fight each other', that democracies have established a 'separate peace'. Forsythe (1992), while accepting this conclusion, has a different main concern: to investigate a semi-deviant case, how democracies may substitute covert action for overt force against popularly elected governments which pursue policies strongly disliked by the United States or other major democratic powers. Sørensen (1992) accepts the Kantian vision, while wishing to retain some basic insights of neorealism. Russett/Antholis (1992) attempt

---

[1]This article was originally published as a 'Focus On' article in *Journal of Peace Research* 29(4): 369–376, 1992 and served as an introduction to a section with articles by Weede, Forsythe, Sørensen, and Russett/Antholis. The abstract has been added.
[2]I would like to thank Bruce Russett, Anne Julie Semb, Harvey Starr, Erich Weede, and several members of the editorial committee of the *JPR*, particularly Torbjørn L. Knutsen, for excellent comments on an earlier draft. Since I am also the editor of the *JPR*, it is particularly appropriate to emphasize that views expressed in this column are solely the responsibility of the author.

© The Author(s) 2015
N.P. Gleditsch, *Nils Petter Gleditsch: Pioneer in the Analysis of War and Peace*, SpringerBriefs on Pioneers in Science and Practice 29,
DOI 10.1007/978-3-319-03820-9_4

to extend the coverage of the proposition that democracies rarely fight each other to the ancient Greek city-states. Two issues ago, Starr (1992) sought to link relatively recent empirical findings about the lack of war among democracies to more established theoretical ideas about pluralistic security communities (Deutsch et al. 1957).

These are but a few examples from a burgeoning literature. Every recent volume of the leading journals in international relations and peace research contains articles on some major or minor aspect of this theme.

The observation that democracies do not fight each other was made almost three decades ago by Babst (1964, 1972).[3] Babst had examined data on 116 major wars from 1789 to 1941 from Wright (1965) and found that 'no wars have been fought between independent nations with elective governments' (1964: 10). Applying a probabilistic argument to the two world wars of this century, he concluded that it was extremely unlikely all the elective governments (10 out of the 33 independent nations participating in World War I; 14 out of 52 in World War II) should be on the same side purely by chance.

A second round of debate was initiated by Rummel (1983), who argued that 'libertarian' states were more peaceful and that libertarian states never fought each other.[4] His argument quickly led to rejoinders by Chan (1984), Weede (1984), and others. At the same time, Doyle (1983, 1986) was developing an argument based on the views of Kant.

After nearly a decade of debate following Doyle's and Rummel's articles, there is now a near-consensus on two points: that there is little difference in the amount of war participation between democracies and non-democracies (Rummel being the major dissenter here) but that wars (or even military conflicts short of war) are non-existent (or very rare) among democracies. Indeed, several scholars have echoed Levy's statement that this 'absence of war between democratic states comes as close as anything we have to an empirical law in international relations' (Levy 1989: 270). This empirical regularity has never been seriously called into question.

Enthusiasm for this remarkable finding should be tempered with an appreciation that it applies only in cases where a relatively high threshold is set for both 'democracy' and 'war'. Take democracy first: Most scholars have followed (more or less) the criteria carefully specified by Small/Singer (1976): (a) free elections with opposition parties, (b) a minimum suffrage (10 %); and (c) a parliament either in control of the executive or at least enjoying parity with it. Schweller (1992: 240)

---

[3] According to Doyle (1986: 1166) this empirical regularity was noted by Streit (1938: 88, 90–92), but his book appears to have had little fall-out in the academic literature.

[4] Rummel's views had been stated earlier, in vol. 4 (1979) and vol. 5 (1981), in his magnum opus *Understanding Conflict and War*. In 1979 his proposition 16.11 (Joint Freedom) read: 'Libertarian systems mutually preclude violence' (1979: 277) and he cited Babst's work as evidence. But this was merely one out of 33 wide-ranging propositions within a gigantic philosophical scheme summed up later (1981: 279) in his 'Grand Master Principle': Promote freedom with three corollaries, including no. 3: Freedom maximizes peace from violence. The immoderate pretensions of this scheme, along with Rummel's unrelenting liberalism and extremely hawkish views on defense, may have deterred readers from noticing what was in fact the strongest proposition in the series.

applauds Doyle's definition of 'representative government' with the suffrage level raised to at least 30 % (and female suffrage granted within a generation of its initial demand) and further requires the government to be (d) 'internally sovereign over military and foreign affairs' and (e) stable (in existence for at least 3 years). He also adds (f) individual civil rights and—more controversially perhaps—(g) private property and a free-enterprise economy. Attempts to lower the thresholds in empirical studies have not been successful: During the Peloponnesian Wars, 'democracies were slightly more likely to fight one another than to fight any other type of regime' (Russett/Antholis 1992: 424) even though a norm was found to be emerging among democracies against fighting other democracies. Using cross-cultural ethnographic evidence from the Human Relations Area Files, Ember et al. (1992) found more supportive evidence, but the hypothesis had to be substantially revised to be testable on these data.[5] Even in the modern age, lowering the suffrage threshold makes an anomalous case of the British-American war of 1812.[6]

Lowering the threshold for war below the 1,000 battle deaths used in the Correlates of War datasets on international and extrasystemic wars also produces less clear-cut results. Maoz/Abdolali (1989) and Maoz/Russett (1992) have tested propositions about democracy and war on the dataset on 'militarized interstate disputes', also generated within the COW project (Gochman/Maoz 1984). In the latest of these studies, 15 cases of disputes between democracies have to be accounted for. To forestall criticism that war is so rare an event that 'it is difficult to demonstrate the effectiveness of pacifying conditions' (p. 380), Weede extends his study, too, to the militarized disputes dataset. This yields one such dispute between two democracies, Finland and Norway! Weede fails to find this dispute in other comparable datasets and raises questions about the coding scheme of the militarized disputes data. However, there was in fact a dispute between Finland and Norway in 1976–77 (about German NATO forces in Norway) and this discussion also referred to the friendship treaty between Finland and the USSR, which might be invoked in the case of a new threat from Germany. It may or may not be reasonable to characterize this as a 'militarized dispute'; in any case, such incidents are so far from war that it is unreasonable to assume they should be accounted for by the same factors.[7]

'Democracies don't fight each other'—why was such a simple observation not made in the great classical studies of war? Richardson (1960) did not touch this topic at all. Wright (1965) dealt at some length with the relationship between

---

[5]The proposition tested was that internal warfare was lower in political units with widespread political participation (Ember et al. 1992: 9).

[6]According to Small/Singer (1976: 54, n. 8) British suffrage did not exceed 3 % until 1867.

[7]After I wrote this, Weede reported that the incident is coded as having taken place in 1965, but has no further information. Those responsible for the dataset have been unable to supply any clarification, neither have Finnish researchers whom I have consulted on this problem. Similar episodes to the one mentioned from 1976–77 did occur in the 1960s, although not as serious, and one of them could have resulted in this mysterious coding.

democracy and war, but did not comment on the lack of war between democracies.[8] And why, when this striking regularity was noticed in the early 1960s, did it take nearly thirty years before it became widely acknowledged?

One answer to the latter question may be professional jealousy. Babst was a criminologist and generally published in journals which must be regarded as extremely obscure from a peace research point of view. Nevertheless, his second article was spotted by professional students of war. In a frequently quoted article, Small/Singer (1976: 51) lampooned Babst's finding: 'In a less academic enterprise, a recent issue ... prominently featured an analysis that allegedly ends the debate [on regime-type and foreign conflict behavior] forever.' Such a 'seductive proposition' was likely 'to be accepted uncritically by those searching for some ray of hope in the generally bleak picture of contemporary international relations'. Since they found some of Babst's coding rules to be 'invisible' they generated a new dataset from their own Correlates of War project in order to examine Babst's 'superficially credible proposition'.

A second reason for the late acceptance may be methodological. A number of early contributions to the literature confuse the issue of a national-level proposition (democracies are more peaceful) and a dyadic-level proposition (democracies don't fight each other). Babst argued strictly in terms of the latter (as Kant had done, 169 years earlier), but Singer & Small, in their polemic against Babst, set out to demolish in some detail a proposition of the first type, 'the innate peacefulness of the bourgeois democracies'. Thus, Small & Singer really knocked down a straw man. But so persuasive was their article that for a long while no one pursued this lead. When Rummel joined the battle he chose to defend the very thesis that Small & Singer had disconfirmed, that 'libertarian' states were inherently more peaceful. This drew attention away from his second thesis on dyadic peace between libertarian states.

Despite the increasing methodological sophistication of research on international violence, the difference between a main effect and an interaction effect is not always grasped. Moreover, with an increasing number of nations, the idea of conducting research at the dyadic level—where the number of units of analysis is roughly the square of the number of nations—is not especially appealing, either to research directors or funding agencies.

While the concept of an interaction effect may be too complicated, the finding that democracies don't fight each other may also be seen as too simple, even simplistic. In the midst of regression analyses, factor analyses, and numerous other multivariate techniques, the idea that one variable alone is a sufficient (but not necessary) condition for a state of peace in the sense of non-war seems ridiculously naive. For instance, towards the end of their article Small/Singer (1976: 67) did

---

[8]Wright concluded that continuous war undoubtedly favors despotism. The more democracies, therefore, the greater the value of war to the despots. 'The greater the number of sheep, the better hunting for the wolves' (p. 266). In a pure balance of power system, democracy probably cannot survive. However, he also noted that democracies are better suited to fight long wars because they have stronger economies.

admit that they could only find two very marginal cases of 'bourgeois democracies' fighting each other, but they dismissed this 'superficial proof of the innate peacefulness of the bourgeois democracies' with a comment that this might perhaps be accounted for by geographical proximity, since wars tended to be fought between neighbors and few democracies had common borders. This is only the first of many attempts to explain away the idea of a separate peace among democracies with reference to third variables—many of which have recently been put to rest by Maoz/Russett (1992). Testing for third variables brings the issue into the normal grind of research practice, but such tests for 'statistical artifacts' are less meaningful in the case of perfect or near-perfect correlations.[9] Regardless of the variables used to subdivide a population of nation-pairs, a zero in the cell for joint democracy will remain zero in any subdivision. It is possible, of course, that a third variable might be found which would account for joint democracy as well as for nonwar. But the only third variables which could perform such a feat would themselves need to have a perfect relationship with both the other variables. A little reflection should suffice to show that geographical distance is not such a variable: most wars have been between neighbors, but certainly not all. And although most democracies are not neighbors, some are. In fact, in more than 25 years of research on 'Correlates of War' no one has come up with any relationship nearly as strong as the dyadic relationship between democracy and nonwar. Therefore, it seems extremely unlikely that such an underlying causal factor will be identified. Of course, it will not be hard to find separate third-variable explanations for each separate peace, deterrence here, distance there, and so on. Such explanations will be advanced with particular fervor by those hostile to any quantitative analysis.[10] In fact, by their very diversity they do little to bolster the many armchair generalizations, frequently single-factor ones, about the war-making of democracies.

As far as third variables are concerned, the perfect or near-perfect correlation between democracy and nonwar in dyads should soon begin to have a very different effect: all research on the causes of war in modern times will be regarded as suspect if it is not first corrected for this factor. In fact, I would argue that most behavioral research on conditions for war and peace in the modern world can now be thrown on the scrap-heap of history, and researchers can start all over again on a new basis. Despite mental resistance to such an idea, this is exactly as it should be in a cumulative discipline. A similar caution must be exercised in formulating new hypotheses about war. For instance, a number of authors are currently urging that environmental problems are a major factor in causing war.[11] This general thesis seems extremely implausible if it is meant to include war over environmental issues between democratic countries. As Diehl (1992: 340) points out, some relationships

---

[9]As will be recalled, Maoz/Russett (1992) found a significant number of democratic dyads engaging not in war but in 'militarized disputes'—otherwise their exercise would have been futile.

[10]A good case in point—well-argued in its genre—is Cohen (1991).

[11]For particularly clear examples, see Colinvaux (1980) and Ehrlich/Ehrlich (1972). For more skeptical views see Deudney (1990, 1991) and Lipschutz/Holden (1989).

are so powerful that they supersede any other conditions for war. The proper approach here would be first to sort out the double democratic dyads and then to look at the environmental factors in outbreaks of war in the remaining dyads.[12] Ten years from now, the finding that democratic countries don't war with each other will probably be regarded as extremely trivial in a research design—a factor to be corrected for before we get on with the real job of accounting for the wars that do occur. By then, 'antipositivists' who now reject the democracy-nonwar relationship may revert to their second line of defense, that quantitative research can do nothing but belabor the obvious.

A further reason for discounting the dyadic relationship between democracy and peace was that it seemed to be based, on the one hand, on rather raw empiricism (Babst, Rummel) and, on the other, on airy philosophical principles (Kant, Doyle). Schweller (1992: 235) argues that much of the literature is 'data-driven' and Lake (1992: 24) feels that 'No theory presently exists that can account for this striking empirical regularity'. While this criticism might have been true in the 1970s, it would no longer seem to hold. Fairly elaborate theoretical arguments have been made in terms of constraints on decision-makers in democracies, in terms of democracy as an exercise in non-violent domestic conflict resolution which can be extended to international affairs if a suitable (i.e. democratic) counterpart is found, in terms of democracies seeing the mutual relationships as positive- sum rather than zero-sum, or in terms of state rent-seeking, which creates an imperialist bias in a country's foreign policy, but less so in democracies.[13] While there is as yet no consensus on which theoretical rationale accounts best for the observed relationship—or on how to separate them empirically—there is at least no lack of convincing theories.

The apparent discrepancy between findings at the nation level and at the dyadic level also calls for explanation. Theoretically, of course, the two can easily be linked: the simplest way to do so is to assume that non-democratic nations tend to attack peaceful democratic nations and that the wars fought by democratic countries are always defensive. In this way, the war participation of the democracies becomes as high as that of the non-democracies, even though the former are more peaceful.

---

[12]Like Weede and many others, but contrary to Rummel, I refrain from concluding that the democratic peace is a deterministic relationship, thus making it possible for a single contrary case to falsify the relationship. The various points made here hold even if wars between democracies are only extremely rare and not zero. If there are deviant cases, however, it makes sense to look for third variables to account for those cases.

[13]Lake (1992: 24) conceives of the state as a profit-maximizing firm trading services (mainly protection) for revenues. Autocratic states exhort exorbitant rents at the expense of their societies and therefore tend towards imperialism.

Small/Singer (1976: 66), however, found no differences between democratic and non-democratic countries with respect to war initiation. More recently, Schweller (1992: 249) hypothesized 'that only authoritarian regimes initiate preventive war and that they do so regardless of whether the challenger is democratic or authoritarian', he found the empirical evidence to be 'overwhelming' from Sparta to Nazi Germany.[14] Declining democratic leaders tend to seek accommodation when faced by democratic challengers, and a defensive alliance when challenged by a nondemocracy. Schweller found Israel to be the leading candidate for a deviant case from the latter regularity—one, however, which does not contradict the main regularity that we are discussing here. Morgan/Schwebach (1992: 312) also concluded 'that democratic states are less likely to escalate disputes than are non-democracies'. Lake (1992: 30), on the other hand, maintains that 'democracies are not only less likely to wage war with each other', but that 'they are also significantly more likely to win the wars they fight against autocracies', a regularity which no doubt has contributed to skepticism about the peacefulness of democracies. The debate on this and other theoretical issues will obviously continue—hopefully some of it will take place in this journal!

A fifth reason for not taking much account of the democracy-war relationship at the dyadic level is that when it was first proposed by Kant there were only three liberal regimes in existence (Switzerland, France, and the USA; Doyle 1986: 1164). Thus, Kant's writings might be dismissed as theoretical speculation about a hypothetical future world with no empirical evidence and without much consequence in a world of despots. In the two centuries since then a 'separate peace' has spread to an increasing number of states: roughly 50 for the period since 1945, according to Doyle. Not only are there more democracies around, but their numbers are increasing. When 10 % of the world's nations were democracies (roughly the state of affairs in the 19th century) only close to 1 % of all nation-pairs were excluded from war.[15] With 50 % democracies—not an unrealistic target for the close of this century—the separate peace encompasses close to 25 % of all pairs. This, then, is the basis for the 'obsolescence of war in the developed world' heralded by Mueller (1989).

And finally, a more political explanation for the tardy response of the research community to the idea of the separate democratic peace: Virtually all systematic research concerned with causes of war has taken place in countries affected by the

---

[14]His systematic database included great-power preventive wars from 1665.

[15]Actually, because nations do not engage in wars with themselves, the correct percentage is $(25x-100)/(x-1)$, where x is the number of nations in the international system. As x increases, this comes very close to 25 %. For instance, with 180 nations in the state system (a reasonable description of the present system, although there are some ambiguous cases) the percentage is 24.6. If we also assume that democratic nations are unlikely to engage in civil wars, then the percentage of pairs excluded from war in a world with 10 % democratic nations is exactly 1 %. (And, more generally, y% democratic countries yields y square % pairs excluded from war.)

Cold War. Research attributing major importance to political democracy seemed propagandistic to many peace researchers who subscribed to a 'third way' in the Cold War and disliked anything that smacked of one-sided propaganda for 'the free world'.[16] Babst's original article was not entirely free of this preaching when it suggested that democracy was a great force for peace and that 'diplomatic efforts at war prevention might well be directed toward further accelerating' the growth of elective governments. Small/Singer (1976: 51, n. 3) suggested, however, that Babst's prescription 'could, paradoxically enough, turn out to be a major stimulus to war', an observation compatible with at least some of the rhetoric in the 1991 Gulf War. Among the potentially important policy implications of Rummel's work on this topic, Vincent (1987: 104) singled out one he clearly regarded as unsavory: 'that American covert and overt interventions for the purpose of democratizing a society would help promote peace in the world system'. The debate about imperialism in the 1970s focused, unsurprisingly, more on the war-mongering nature of several democracies than on their peacefulness. But the idea of a democratic separate peace seemed too soft for the realists, who felt more comfortable with deterrence and strict bipolarity (and still do, as is evidenced by the doomsday predictions for Europe after the Cold War in a celebrated article by Mearsheimer 1990).[17] As a former member of the deterrence school of thought, Weede has taken a bold step in reconsidering his own views. Peace researchers who rejected the link between democracy and peace from a radically different paradigm should not be less forthright. The strong finding about the 'democratic peace' may to some extent have been a victim of the Cold War. No wonder then that it fell to an 'innocent criminologist' to observe that the emperors had no clothes. The Cold War has now ended in the real world; it should end in peace research, too.

---

[16]Of course, many countries in the 'free world' were neither free nor peaceful.

[17]A third group which may be reluctant for political reasons to acknowledge the persistence of the dyadic relationship between democracy and peace is the functional (or integrationist) school of thought. Conventional wisdom has it that the creation of the Common Market has helped preserve the peace between the traditional enemies Germany and France. Although the idea of the impossibility of war between highly interdependent countries—put forward with much fanfare by Angell (1910)—should have been thoroughly discredited by World War I, it continues to have strong backing in political thinking on both sides of the Atlantic. If the idea of a separate peace between independent democracies holds, then it has no direct bearing on war and peace if these countries continue in the present European Community, develop it into a European Union, or leave it altogether. Neither does it have any significance if additional democratic countries join a European Union or not—although this may be a good (or a bad) idea, for a number of other reasons. On the other hand, if countries with fragile democracies are allowed to join the European Community and if membership in the EC helps to stabilize their democratic government—two big ifs!—then the European Community may nevertheless function as a peace factor. The same argument can be applied to postwar Germany, Italy, and possibly other European states.

References

Angell, Norman, 1935: *The Great Illusion* (London: Heinemann). [First published in 1910.]

Babst, Dean V., 1964: "Elective Governments—A Force For Peace", in: *Wisconsin Sociologist*, 3,1: 9–14.

Babst, Dean V., 1972: "A Force for Peace", in: *Industrial Research*, 14(April): 55–58.

Chan, Steve, 1984: "Mirror, Mirror on the Wall … Are the Freer Countries More Pacific?", in: *Journal of Conflict Resolution*, 28,4: 617–648.

Cohen, Raymond, 1991: *Apples, Oranges and Lemons: An Appraisal of the Theory that Democracies Do Not Fight Each Other. Paper Presented to the Annual Conference of the British International Studies Association* (University of Warwick): 16–18 December.

Colinvaux, Paul, 1980: *Fates of Nations: A Biological Theory of History* (New York: Simon & Schuster).

Deudney, Daniel, 1990: "The Case Against Linking Environmental Degradation and National Security", in: *Millennium*, 19,3: 461–476.

Deudney, Daniel, 1991: "Environment and Security: Muddled Thinking", in: *Bulletin of the Atomic Scientists*, 47,3: 22–28.

Deutsch, Karl W.; Burrell, Sidney A.; Kann, Robert A.; Lee, Maurice, Jr.; Lichtermann, Martin; Lindgren, Raymond E.; Loewenheim, Francis L.; Van Wagenen, Richard W., 1957: *Political Community and the North Atlantic Area* (Princeton, NJ: Princeton University Press). [For an extensive extract, see pp. 1–91 in (1966) *International Political Communities. An Anthology* (Garden City, NY: Doubleday).]

Diehl, Paul, 1992: "What Are They Fighting For? The Importance of Issues in International Conflict Research", in: *Journal of Peace Research*, 29,3: 333–344.

Doyle, Michael W., 1983: "Kant, Liberal Legacies, and Foreign Affairs. Part 1", in: *Philosophy & Public Affairs*, 12,3: 205–235; part 2, 12,4: 323–353.

Doyle, Michael W., 1986: "Liberalism and World Politics", in: *American Political Science Review*, 80,4: 1151–1169.

Ehrlich, Paul R.; Ehrlich, Anne H., 1972: *Population, Resources, Environment. Issues in Human Ecology* (San Francisco, CA: Freeman).

Ember, Carol R.; Ember, Melvin; Russett, Bruce, 1992: "Peace Between Participatory Polities: A Cross Cultural Test of the 'Democracies Rarely Fight Each Other' Hypothesis", in: *World Politics*, 44,4: 573–599.

Forsythe, David, 1992: "Democracy, War, and Covert Action", in: *Journal of Peace Research*, 29,4: 385–395.

Gochman, Charles S.; Maoz, Zeev, 1984: "Militarized Interstate Disputes 1816–1976", in: *Journal of Conflict Resolution*, 28,4: 585–616.

Kant, Immanuel, 1795: "*Zum ewigen Frieden* [Perpetual Peace]". English version, in: Reiss, Hans (Ed.) *Kant's Political Writings* (Cambridge: Cambridge University Press) 1970.

Lake, David A., 1992: "Powerful Pacifists: Democratic States and War", in: *American Political Science Review*, 86,1: 24–37.

Levy, Jack S., 1989: "The Causes of War: A Review of Theories and Evidence", in: Tetlock, Philip E.; Husbands, Jo L.; Jervis, Robert; Stern, Paul C.; Tilly, Charles (Eds.): *Behavior, Society, and Nuclear War* (New York: Oxford University Press): Vol. 1, 209–313.

Lipschutz, Ronnie D.; Holdren, John P., 1990: "Crossing Borders: Resource Flows, the Global Environment, and International Security", in: *Bulletin of Peace Proposals*, 21,2: 121–133.

Maoz, Zeev; Abdolali, Nasrin, 1989: "Regime Types and International Conflict, 1816–1976", in: *Journal of Conflict Resolution*, 33,1: 3–35.

Maoz, Zeev; Russett, Bruce, 1992: "Alliances, Contiguity, Wealth, and Political Stability: Is the Lack of Conflict Among Democracies a Statistical Artifact?", in: *International Interactions*, 17,3: 245–267.

Mearsheimer, John J., 1990: "Back to the Future: Instability in Europe After the Cold War", in: *International Security*, 15,1: 5–56. [See also comments by Stanley Hoffmann, Robert O.

Keohane, Thomas Risse-Kappen, and particularly Bruce M. Russett and responses from John J. Mearsheimer in the subsequent issues.]

Morgan, T. Clifton; Schwebach, Valerie L., 1992: "Take Two Democracies and Call Me in the Morning", in: *International Interactions*, 17,4: 305–320.

Mueller, John, 1989: *Retreat from Doomsday. The Obsolescence of Major War* (New York: Basic Books).

Richardson, Lewis Fry, 1960: *Statistics of Deadly Quarrels* (Pittsburgh, PA: Boxwood/Chicago, IL: Quadrangle).

Rummel, Rudolph J., 1979: *War, Power, Peace. Vol. 4 of Understanding Conflict and War* (Beverly Hills, CA & London: Sage).

Rummel, Rudolph J., 1981: *The Just Peace. Vol. 5 of Understanding Conflict and War* (Beverly Hills, CA & London: Sage).

Rummel, Rudolph J., 1983: "Libertarianism and International Violence", in: *Journal of Conflict Resolution*, 27,1: 27–71.

Russett, Bruce; Antholis, William, 1992: "Do Democracies Fight Each Other? Evidence from the Peloponnesian War", in: *Journal of Peace Research*, 29,4: 415–434.

Schweller, Randall L., 1992: "Domestic Structure and Preventive War. Are Democracies More Pacific?", in: *World Politics*, 44,2: 235–269.

Small, Melvin; Singer, J. David, 1976: "The War-Proneness of Democratic Regimes", in: *Jerusalem Journal of International Relations*, 1,4: 50–69.

Starr, Harvey, 1992: "Democracy and War: Choice, Learning, and Security Communities", in: *Journal of Peace Research*, 29,2: 207–213.

Streit, Clarence, 1938: *Union Now: A Proposal for a Federal Union of the Leading Democracies* (New York: Harpers).

Sørensen, Georg, 1992: "Kant and Processes of Democratization: Consequences for Neorealist Thought", in: *Journal of Peace Research*, 29,4: 397–414.

Vincent, Jack, 1987: "Freedom and International Conflict: Another Look", in: *International Studies Quarterly*, 31,1: 103–112. [See also a comment by Rummel on pp. 113–117 and a rejoinder by Vincent on pp. 119–126.]

Weede, Erich, 1984: "Democracy and War Involvement", in: *Journal of Conflict Resolution*, 28,4: 649–664.

Weede, Erich, 1989: "Extended Deterrence, Superpower Control, and Militarized Interstate Disputes", in: *Journal of Peace Research*, 26,1: 7–17.

Weede, Erich, 1992: "Some Simple Calculations on Democracy and War Involvement", in: *Journal of Peace Research*, 29,4: 377–383.

Wright, Quincy, 1965: *A Study of War. Second Edition with a Commentary on War Since 1942* (Chicago, IL & London: University of Chicago Press). [First edition published in 1942.]

Chapter 5
The Treholt Case

The spy charges and court case against Arne Treholt, a Norwegian civil servant and politician, have led to a long-standing controversy in the Norwegian media.[1] This article examines the literature on the Treholt case for the first ten years after his arrest. The literature is classified under a scheme borrowed from Cold War history: traditionalist ('Arne Treholt was the greatest spy ever caught in Norway'), revisionist ('Treholt was a victim of a political vendetta against the left'), and post-revisionism (studies using greater historical distance and declassified archives to gain a more independent perspective – and it remains to be seen what the conclusion will be). Most of the published literature is revisionist, but none of the books classified under this heading offer a fully satisfactory answer to the traditionalist account. In particular, the revisionist literature fails to explain plausibly why Treholt and his case officers engaged in so much risky and covert behavior if all they did was to exchange political views. The effects of Treholt's espionage in the sense of traditional national security may have been overestimated; his value as a political informer was probably far greater.

5.1 Introduction

The Treholt case is, arguably, the most serious *political* spy case to be revealed in the West after the Second World War. Of course, in several NATO countries spies within the military establishment and in the intelligence agencies have had access to more sensitive national security material than Treholt in Norway and the opportunity to pass along greater volumes of it. But one is hard put to name a spy with a higher *political* position than Arne Treholt, who served as a state secretary (the second highest political position) in Norway's Ministry of Ocean Law, and before

---

[1]This article was originally published in *Intelligence and National Security* 10(3): 529–538. The abstract has been added. The past 20 years have seen a large increase in writings on the Treholt case, few of which add materially to the substance. The remaining portions of the verdict have now been declassified.

© The Author(s) 2015
N.P. Gleditsch, *Nils Petter Gleditsch: Pioneer in the Analysis of War and Peace*, SpringerBriefs on Pioneers in Science and Practice 29,
DOI 10.1007/978-3-319-03820-9_5

that as Political Secretary to the same minister, then Minister of Commerce and, as such, a member of the exclusive Security Committee of the Norwegian Cabinet. In 1978 the tasks of the Ministry of Ocean Law were completed, the Ministry was incorporated into the Foreign Ministry, and Treholt—then already on the short list of suspected spies—became a regular Foreign Service officer, although he continued to play an informal role in Labor Party politics. Treholt held higher formal office than Günther Guillaume, although the politician to which he had attached his career, Jens Evensen, was not as centrally placed as Willy Brandt and the country in which he operated was of course much smaller and more peripheral than West Germany. Nevertheless, he has been characterized by Oleg Gordievsky as one of the KGB's ten most important agents.

In Norway the Treholt case came as a big shock when revealed to the public on 21 January 1984, the day after his arrest at Oslo airport while on his way to a meeting in Vienna with his case officer, KGB General Gennadi Titov. The case has remained embroiled in political controversy, much like the Rosenberg case in the USA or the Petrov case in Australia.

The first book with Treholt's picture on the front cover was published just a few months after his arrest. His wife published a revealing book about a year later, before the trial started, and by the summer of 1994 my count had reached 17 non-fiction books dealing exclusively with the Treholt case. Add to this two slightly disguised novels on the case, and a half a dozen non-fiction books where the Treholt case played a major role. In addition, Norwegian law journals have published relevant articles, and there have been two debates in Parliament, one of them preceded by a lengthy committee report on why Treholt was admitted as a student at the Norwegian National Defense College. I have provided an extensive bibliography in an article for a Norwegian journal, along with a more detailed discussion of the literature[2] and will limit myself here to the more important contributions to the debate, as well as the few contributions available in English.

The Treholt case belongs to the Cold War and it seems appropriate to interpret it within the framework of the tri-partite division of the literature on the Cold War: There is, first, the *traditionalist* school, an explanation of and to a large measure an apology for the official Western position. Then there is the *revisionist* school of critical studies, frequently coming close to an apology for the other side, and finally, there is the *postrevisionist* school, with a diverse set of studies using greater historical distance and recently opened archives to gain a more independent perspective.

---

[2]Gleditsch (1994).

5.2 Traditionalism

The two most important works in the traditionalist literature are the verdict (Eidsivating Lagmannsrett 1985)[3] and the volume by Michael Grundt Spang.[4] The lower court verdict (which eventually became final) was published as a 250-page book and there is also an English translation—although it is not as readily accessible.[5] Michael Grundt Spang, a senior journalist in Norway's largest newspaper *VG*, has based his account on the extremely detailed reports of his paper. Since verbatim court records are not available from Norwegian trials, the *VG* reports and other news reporting, will remain the major source of what was said at the trial.[6] Spang's views are close to views of the prosecutor, 'a betrayal of dizzying proportions', and his conclusion close to that of the court which imposed the maximum of penalty of 20 years. Treholt was convicted of handing over to KGB representatives a great deal of information to which he had access in the line of duty, and which was judged to be detrimental to national security. A great deal of this information he acquired while a student at the Defense College, which he attended as an official of the Foreign Ministry. In addition, he was also convicted for espionage for Iraq.

Other important books in the same tradition are the memoirs of Haarstad (1988), head of the Norwegian Special Branch for most of the seven-year period when Treholt was under investigation, and Tofte (1987), the Special Branch veteran who led the investigation, made the arrest, and conducted most of the interrogation. The two books of Treholt's (now former) wife, which provide ample documentation of how Treholt kept family and friends completely in the dark, must be included in the same category.

The most troublesome point in the traditionalist literature concerns the proof for the actual transmission of sensitive information. Although Treholt initially confessed to being caught in a trap by the KGB (he later retracted this) he never admitted having handed over national security information. Broadly speaking, the stronger the evidence for transmission—or attempted transmission, the less significance for national security. For example, the 66 documents found in Treholt's briefcase when he was arrested, were judged by the court *not* to be relevant to national security, either separately or as a whole.[7]

---

[3]Further material from the verdict was declassified on 23 August 1991 and is available from the Norwegian Special Branch, Politiets Overvåkingstjeneste.

[4]Spang (1986).

[5]I obtained my copy from the Norwegian Foreign Ministry.

[6]Extensive records of court sessions in early 1986 recording evidence for the (later aborted) Supreme Court review of the case are, however, available for inspection in the Special Branch.

[7]The Norwegian Supreme Court has laid down a doctrine ('the combination principle' or 'the puzzle principle') whereby pieces of information which separately are not judged to have national security implications may have such implications if revealed together. This doctrine, when applied to information revealed publicly rather than transmitted to case officers in secret, has some troubling implications for freedom of expression. Cf. Gleditsch (1987) and Gleditsch/Wolland (1991).

In the end, the verdict relied heavily on the damage done by the transmission to the KGB of information Treholt acquired at the Defense College. By the time he was admitted to the Defense College he was under very strong suspicion, although definitive evidence to convict him was still lacking. Thus, the question arises whether the Norwegian authorities should not have prevented his term at the Defense College. Sending Treholt to the Defense College was the idea of a Foreign Ministry colleague in charge of finding suitable entrants who was not cleared for the information about Treholt's probable espionage.

Another weakness (more from a historical than from a legal point of view) of the traditionalist literature is that it assumes that Treholt needed a motive for espionage, but fails to establish conclusive what that motive might be. Initially, the prosecution placed great emphasis on Treholt's own confession (the one he had later retracted) which told in detail how as a student political activist he became involved in a relationship which he lost control of. Towards the end of the case, the profit motive came to the fore and the verdict explicitly rejected the idea that he was pressured by the Russians. These two motives might, of course, apply to different phases of his secret life. There was never any suggestion of pressure in the case of his espionage for Iraq, and this belongs to the late phase. The uncertainty about motives derives in great measure from Treholt's retraction of his first statements to the police. By insisting that he was acting freely at all times, and insisting on full acquittal, Treholt and his lawyers awarded a walkover victory to the prosecution. However, for anyone who attempts to comprehend why the career of a talented and ambitious young politician ended in such a tragedy, the traditionalist literature seems ambiguous and incomplete. The second book by his estranged wife reveals that one of the FBI officers assigned to his case (posing as a friendly neighbor in Manhattan) thought that he was a communist; such ideological motives find little support in the Norwegian literature.

5.3 The Revisionist Critique

In the dozen revisionist works it is not simple to point to the main contributions. Jo Bech-Karlsen[8] covered the court case for a left-wing weekly and his book is a good summary which includes the many second thoughts and question-marks that many radicals had about the case, once the initial shock (and condemnation) had receded into the background. Treholt's own book[9] about the arrest and interrogation is a literary masterpiece, which received a prize in a non-fiction literature competition. His argument is so self-serving, however, and some of his excuses so transparent that the book might have been better submitted in a competition for fiction. A very

---

[8]Bech-Karlsen (1985).
[9]Treholt (1985).

interesting book by Calmeyer (1993) about political surveillance in the Norwegian labor movement attempts to link this to the Treholt case, but not very successfully.

A separate strand of revisionist literature deals with the pre-trial publicity. This literature provides extensive documentation that there was a massive pre-trial condemnation of Treholt in quasi-legal terms from politicians and the media, with Prime Minister Kåre Willoch as the most prominent example. This has led to a great deal of self-criticism in the journalistic profession, and to a lesser extent among politicians. However, it is impossible to assess the significance of this publicity for the verdict. Treholt's status as a media star (even before his arrest) clearly works both ways. Quite apart from the pre-trial condemnation, he has also had a very active circle of friends who have argued his case publicly, a degree of public support not enjoyed by any other Norwegian charged with espionage.

The mainstream revisionist critique of the verdict relates to lack of concrete proof for the transmission of national security information, as well as the unfortunate admission of Treholt to the Defense College. Some go beyond this to argue that his term at the Defense College was a deliberate trap that the government set in order to obtain sufficient evidence to convict him. This question was extensively investigated by Parliament and the weight of the evidence clearly indicates that his admission was a bureaucratic error, although there is disagreement about how the error might have been prevented, by whom, and at what stage.

Several law professors have criticized various aspects of the investigation and the trial, notably the legality of a secret raid on Treholt's apartment in Oslo, on the interpretation of the relevant paragraphs in the penal code, on the question of intent, etc. To a non-lawyer like myself, much of this criticism seems sufficiently well-founded to make it a reasonable expectation that the Supreme Court would have reduced the sentence, perhaps significantly. Treholt actively prevented this, by with-drawing his appeal after the Supreme Court proceedings had started (as a reaction to some procedural decisions of the court which went against him). Instead, he put his trust in a full new review on the lower court (a strategy which was unsuccessful) and eventually in asking for a pardon (finally granted on 3 July 1992). Some revisionists consider that the Court should have limited the verdict to the Civil Service Act (which Treholt clearly had violated), such violations have an upper limit of three years; while others portray him merely as an unconventional diplomat carrying out the kind of private diplomacy and openness which has been publicly applauded after the demise of the East-West conflict. Calmeyer, among others, interprets the persecution of Treholt as the culmination of a long struggle for power within the Labor Party, through which the establishment was finally able to discredit the left wing.

A detailed discussion of the revisionist position would easily lead to a new book. However, a couple of frequently made points are worth mentioning. First, it has become a favorite theme for the revisionists that Treholt cannot be a major spy because he was photographed with his case officers in restaurants. Given how frequently spies have met their case officers in restaurants and the origin of that covertly obtained picture (taken apparently, by Norwegian Special Branch operatives using a camera hidden in a baby carriage) this argument does not have much force. Treholt's extensive 'home archive' has also been adduced as proof that he cannot have been much of a spy because he failed to take elementary precautions. This argument overlooks the most important aspect of his home archive: that it does not contain any notes from his meetings with his case officers, even though Treholt was a compulsive note-taker. His calendar entries about the time for next meetings were carefully coded, although the Special Branch broke the code after they had covertly obtained copies of his calendars. This ultimate consequence of this line of argument is that no one can be a major spy if he is ever caught. Or in other words: If there is no evidence of espionage, the defendant must of course be acquitted. If there is evidence, he is unprofessional and therefore not a major spy, and therefore must also be acquitted.

The charge that the Treholt case was a politically manipulated move, relies on a supposition that other leftist politicians behaved in the same way as Treholt. No one, however, has been able to point to another left-wing politicians who conducted secret meetings with senior KGB officials (one of whom had been publicly expelled from Norway because of his involvement in an earlier Foreign Ministry spy case) behind the back of his friends, family, and colleagues. In fact, the authorities were careful to stress the uniqueness of the Treholt case and the faint beginnings of a political witchhunt were squashed very quickly. Thus, the Treholt case is very different from the British saga of the Cambridge ring, with its endless quest for the third man, etc. When a few people on the Left have equated Treholt's modus operandi with standard political practice on the Left, they have invited a criminalization of their own political activity—an invitation their political opponents, not to speak of the prosecuting authorities, have declined to accept.[10]

The greatest problem with the revisionist literature, however, is that it contains no plausible explanation of why Treholt and General Titov should take the trouble, risk, and expense of meeting covertly in Helsinki and Vienna if all they ever did was to exchange political generalities which Treholt never even bothered to record or pass on to anyone. While the revisionists have a point when they criticize the court for not entertaining rival interpretations to the espionage hypothesis, they themselves have been unable to suggest any plausible hypothesis for Treholt's behavior, or for Titov's.

---

[10]Shawcross (1986) speculates that 'the politics of the time', particularly anti-Americanism was an important motive behind Treholt's espionage. He cites an anonymous Norwegian official to the effect that 'If Treholt could do it, how many more like him are there who grew up in his generation, not only in Norway but in all of Europe?' Eight years later there is a great deal of political debate about the politics of the Left during the Cold War, but little evidence for the idea that this created a generation of spies.

5.4 A Post-revisionist Synthesis?

By and large, the existing literature is strongly polarized and rather unsatisfactory. The revisionist writings are the most voluminous and also contain the strongest exaggerations. This proves nothing about the case itself but illustrates the common phenomenon that while the establishment rules, the opposition argues. There is an asymmetry in exaggeration: the establishment rules to excess, by sending Treholt to jail for an unprecedentedly long period and on somewhat shaky evidence, the opposition engages in conspiracy theories and other strained theorizing.

Norwegian historians, now actively involved in several studies of the connections between the intelligence services and political surveillance in the postwar period, have not yet tackled the Treholt case. Short treatments in a popular history of the postwar period by radio journalist Yngvar Ustvedt (1992) as well as the entry in the leading Norwegian encyclopedia *Store norske konversasjonsleksikon* (1987) have a mildly revisionist flavor. Apart from a book by Arne Treholt's father, a former Cabinet Minister,[11] politicians' memoirs have not yet shed any light on the case. Kåre Willoch's half-page treatment of the Treholt case in his volume from his 5½ years as Prime Minister[12] (when the decision was made to let Treholt into the Defense College, and later to arrest him) is reminiscent of Harold Wilson's cursory treatment of the secret services in his memoirs. More recently, a biography of Thorvald Stoltenberg,[13] a close colleague in the Labor Party, dealt more extensively with the case—but without providing any significant new information.

Excellent post-revisionist works exist for the politically controversial Rosenberg of 1946 case[14] and the equally controversial Petrov case of 1955 in Australia.[15] Both of these books made extensive use of previously unavailable archival material. While the main positions of the revisionist *critiques*—that the Rosenbergs were not spies and that the Petrov case was a plot to discredit the Australian Labour Party— were refuted in these works, the traditionalist position was modified in very significant ways, What are the prospects for a post-revisionist reinterpretation of the Treholt case?

To some extent the future is already here. The volume on the KGB by Christopher Andrew and Oleg Gordievsky published in 1990[16] provided the first confirmation from inside the KGB of Treholt's perceived value as a spy. Some of this information had been revealed in a secret session in the Treholt trial, but it was then presented in extremely veiled terms because Gordievsky was still in place. The book does not provide much new information on the case (in fact it relies mostly on Tofte's memoirs and does not even cite the more extensive and authoritative court

---

[11]Treholt/Hegge (1989).

[12]Willoch (1990).

[13]Salvesen (1994).

[14]Radosh/Milton (1983).

[15]Manne (1987).

[16]Andrew/Gordievsky (1990).

verdict); the force lies in Gordievsky's personal observations and his authority as an official of the KGB department dealing with Scandinavian affairs. Two years later, another KGB officer, Mikhail Butkov,[17] who had left the sinking ship in 1991, published a volume of memoirs augmenting Gordievsky's information. While Butkov was much less senior, he had been stationed in Oslo after the Treholt trial and confirms that official policy of the KGB was to deny Treholt's status as an agent and that it was seen as a publicity victory for the KGB when Calmeyer published an interview with Titov in *Arbeiderbladet* in 1990. It may safely be assumed that journalists of all persuasions have conducted a fairly intensive search for former KGB officials willing to talk about the Treholt case. Apart from the rather unconvincing interview with Titov (and another interview with Treholt's case officer in during the UN period, Vladimir Zhizhin, published in *Aftenposten* in 1993, in which he refused to talk about the Treholt case) all such revelations to date have gone in Treholt's disfavor.

Statements by former KGB officials about Treholt's significance as an agent are very destructive for the revisionist interpretation, but do not necessarily provide any support for the traditionalists. The ex-KGB'niks have provided no new documentation of Treholt revelations which concern national security. In fact, Gordievsky has emphasized Treholt's role as an agent of influence and as a collector of political intelligence.[18] But Treholt was not charged with such offenses, and apart from violating his obligations as a civil servant it is very hard to see what charges they could have led to. If there had been concrete evidence, for instance, of leaks from Treholt of Norwegian negotiating positions during the talks on the delimitation of the economic zone on the Barents Sea (in which Treholt played a key role and which ended in a very controversial agreement) the political condemnation would have been massive, but it is not obvious that this would have been a chargeable offence under the espionage legislation. The prosecution did make an attempt to extend the concept of a secret as described in the penal code, but this attempt was largely rejected by the court, in line with the argument by a minority of the court-appointed experts.

In August 1991 the Court declassified information from 18 pages of the secret part of the Treholt verdict. Even though the new information does in my view add to the evidence for his technical guilt under these particular paragraphs, still, approximately 100 pages of the verdict remain wholly or partly classified. Treholt's defense lawyer has argued that increased openness about the verdict and the court proceedings will speak to the defendant's favor. This may be correct in the limited sense that some of the mystique about the closed sessions evaporates. However, so far Treholt has not had much more success with either Norwegian or Russian *glasnost'* than the sons of the Rosenbergs have had with Khrushchev's unexpurgated memoirs, Sadupletov's memoirs, or and other new information on the Rosenberg case.

---

[17]Butkov (1992). As far as I have been able to ascertain, Butkov has not published anything in English.

[18]Andrew/Gordievsky (1990: 476) and my interview with Gordievsky in London, 30 March 1992.

An uncertain factor is whether the authorities possessed any information which they refrained from presenting in court. The fact that they revealed some of Gordievsky's information (even though it occurred in a super-secret session and though he was not named, he must have been exposed to considerable danger) and presented the illegally obtained evidence from the search of Treholt's apartment in court, indicates that the prosecution was hard pressed to make its case stick. However, if the authorities had had access to signals intelligence, it seems very likely that it would not have been kept outside the courtroom, in order not to endanger 'sources and methods'. We know that signals intelligence—the Venona material—played an important role behind the scenes in the Rosenberg case. A slight preview of a possible future development in the Treholt case was provided by Ørnulf Tofte, who is cited by Treholt as having said in an early interrogation that there was a significant increase in radio traffic from the Soviet embassy in 1973–74, that is, several years after Treholt was recruited and when he was first employed in an official position (as politically secretary to Jens Evensen during the negotiations for a trade agreement between Norway and the Common Market). It is not known whether reference was made to such traffic analyses in closed sessions of the trial, but it would in any case be a large step from that to reveal the contents of decrypted intelligence, if any such exists.

In spy cases prosecuting authorities have been known to cut corners. They represent the collective anger of the nation over treasonous activities, they 'know' that the defendant lies through his teeth, they may know more than they can reveal to the court, technical guilt is notoriously hard to establish, and the secrecy of the proceedings provide a golden opportunity for getting away with more than would be possible in other criminal cases. When, as in the Rosenberg and Treholt trials, they get free points from the defense lawyers by unrealistic insistence on total acquittal, the outcome may turn out to be unduly harsh. My tentative guess about a post-revisionist literature on the Treholt case is that he may be seen to have provided the Russians with much more information than included in the charges, while the legal basis for his conviction under the national security paragraphs of the penal code was nevertheless weaker than the verdict made out. The drastic conclusion here may be that an involuntary alliance of Cold War crusaders on both sides sent Treholt to prison for much longer than would have been justified, had the complete evidence been publicly available, but that he may have been let off too lightly in the political arena.

When the Norwegian book market is overloaded with Treholt books, the available English-language information is remarkably sparse.[19] So while waiting for the post-revisionist revelations, a solid journalistic book in English could serve a useful function.

---

[19]A short early summary of the case can be found in Ausland (1984). A more recent assessment is Shawcross (1988). A short treatment of the Treholt case in Brook-Shepherd (1988: 275) contains several major errors. There must have been a fair amount of coverage in English-language newspapers at the time, but searches on *National Newspaper Index* and *Magazine Index* on Dialog yielded only 16 and 4 items respectively. Apart from the information attributed to Gordievsky in his joint book with Andrew, I am not aware of any first-hand material in the English-language literature on the case.

References

Andrew, Christopher; Gordievsky, Oleg, 1990: *KGB. The Inside Story of Its Foreign Operations from Lenin to Gorbachev* (New York: HarperCollins/London: Hodder).

Ausland, John, 1984: "A Mole Unmasked: The Arne Treholt Spy Case", in: *The Norseman*, 2: 55–57.

Bech-Karlsen, Jo, 1985: *Bange anelser. En bok om Treholt-saken* [Anxious Premonitions. A Book about the Treholt Case] (Oslo: Gyldendal).

Brook-Shepherd, Gordon, 1988: *The Storm Birds: Soviet Post-War Defectors* (London: Weidenfeld).

Butkov, Mikhail V., 1992: *KGB i Norge – det siste kapittel* [KGB in Norway—the Last Chapter] (Oslo: Tiden).

Calmeyer, Bengt, 1993: *Forsinket oppgjør. Arbeiderbevegelsen og den politiske overvåking* [Delayed Reckoning. The Labor Movement and Political Surveillance] (Oslo: Aschehoug).

Eidsivating Lagmannsrett, 1985: *Treholtdommen* [The Treholt Verdict] (Oslo: Norwegian University Press).

Gleditsch, Nils Petter, 1987: "National Security and Freedom of Expression. The Scandinavian Legal Battles", in: *Journal of Media Law and Practice* (June): 2–5.

Gleditsch, Nils Petter, 1994: "Treholt-litteraturen: en studie i polarisering [The Treholt Literature: A Study in Polarization]", in: *Internasjonal Politikk*, 52,2: 275–87.

Gleditsch, Nils Petter; Wolland, Steingrim, 1991: "Norway", in: D'Souza, Frances et al. (Eds.): *Information Freedom and Censorship. World Report 1991* (London: Library Association): 287–291.

Gordievsky, Oleg, 1992: Interviewed by Nils Petter Gleditsch, London, 30 March.

Haarstad, Gunnar, 1988: *I hemmelig tjeneste. Etterretning og overvåking i krig og fred* [In Secret Service, Intelligence and Surveillance in Times of War and Peace] (Oslo: Aschehoug).

Manne, Robert, 1987: *The Petrov Affair: Politics and Espionage* (Sydney: Pergamon).

Radosh, Ronald; Milton, Joyce, 1983: *The Rosenberg File. A Search for the Truth* (New York: Rinehart & Winston/London: Weidenfeld).

Salvesen, Geir, 1994: *Thorvalds verden* [Thorvald's World] (Oslo: Schibsted).

Shawcross, William, 1986: "The New Breed of Anti-US Spy in Europe", in: *Spectator*, 7 June [Reprinted in *Washington Post*, 15 June].

Shawcross, William, 1988: "The Case of the Pampered Spy", in: *Readers Digest* (June): 55–57.

Spang, Michael Grundt, 1986: *Treholtsaken. Hva skjedde?* [The Treholt Case—What Happened?] (Oslo: Aschehoug).

Tofte, Ørnulf, 1987: *Spaneren. Overvåking for rikets sikkerhet* [Undercover Agent. Surveillance for National Security] (Oslo: Gyldendal).

Treholt, Arne, 1985: *Alene* [Alone] (Oslo: Cappelen).

Treholt, Thorstein; Hegge, Oddvar, 1989: *I sol og skygge* [In the Sun and in the Shadows] (Oslo: Gyldendal).

Willoch, Kåre, 1990: *Statsminister* [Prime Minister] (Oslo: Schibsted).

Chapter 6
Armed Conflict and the Environment

Conflict over scarce resources, such as minerals, fish, water, and particularly territory, is a traditional source of armed struggle. Recently, wide-ranging claims have been made to the effect that environmental degradation will increase resource scarcity and therefore contribute to an increase in armed conflict. So far, there has been much controversy and little relevant systematic study of this phenomenon. Most scholarship on the relationship between resources, the environment, and armed conflict suffers from one or more of the following problems: (1) there is a lack of clarity over what is meant by 'environmental conflict'; (2) researchers engage in definitional and polemical exercises rather than analysis; (3) important variables are neglected, notably political and economic factors which have a strong influence on conflict and mediate the influence of resource and environmental factors; (4) some models become so large and complex that they are virtually untestable; (5) cases are selected on values of the dependent variable; (6) the causality of the relationship is reversed; (7) postulated events in the future are cited as empirical evidence; (8) studies fail to distinguish between foreign and domestic conflict; and (9) confusion reigns about the appropriate level of analysis. While no publications are characterized by all of these problems, many have several of them. This article identifies a few lights in the wilderness and briefly outlines a program of research.

6.1 War, Resources, and the Environment

'Nations have often fought to assert or resist control over war materials, energy supplies, land, river basins, sea passages and other key environmental resources.'[1] This passage from the World Commission on Environment and Development

---

[1]This article was originally published in *Journal of Peace Research* 35(3): 381–400, 1998. As can be seen in the bibliography in this volume, no other article of mine has been reprinted so many times.

© The Author(s) 2015
N.P. Gleditsch, *Nils Petter Gleditsch: Pioneer in the Analysis of War and Peace*, SpringerBriefs on Pioneers in Science and Practice 29, DOI 10.1007/978-3-319-03820-9_6

(Brundtland 1987: 290) summarizes a common view of armed conflict.[2] Thus, Renner et al. (1991: 109) claim that 'throughout human history, but particularly since the system of sovereign nation states, struggles over access to and control over natural resources ... have been a root cause of tension and conflict' and that 'history provides numerous examples of how states and nations were destabilized by environmental collapse leading to famine, migration and rebellion'. Galtung (1982: 99) has argued that 'wars are often over resources'. Brock (1991: 409) asserts that 'control over natural resources has always been important in enabling a country to wage war', citing as an example the Pacific War (1879–84) between Chile and Peru over guano deposits. Westing (1986: particularly Chap. 1 and Appendix 2) has examined how resource competition has contributed to the onset of twelve armed confrontations in the 20th century, ranging from the two World Wars to the Anglo–Icelandic 'Cod Wars' of 1972–73. A more ambitious claim is made by Colinvaux (1980: 10), who asserts that 'history has been a long progression of changing ways of life and changing population', with 'wars, trade and empire' as the results. Ehrlich/Ehrlich (1972: 425) argue that 'population-related problems seem to be increasing the probability of triggering a thermonuclear Armageddon'.

Since the emergence of environmental issues on the international political agenda in the early 1970s, there has been increasing concern that environmental disruption is likely to increase the number of disputes originating from competition for scarce resources.[3] Galtung (1982: 99) has argued that 'destruction of the environment may lead to more wars over resources', and suggests that 'environmental effects make a country more offensive because it is more vulnerable to attack and because it may wish to make up for the deficit by extending the eco-cycles abroad, diluting and hiding the pollution, getting access to new resources'. After the end of the Cold War, similar statements have become very common. Opschoor (1989: 137) asserts that 'ecological stress and the consequences thereof may exacerbate tension within and between countries', and Lodgaard (1992: 119) has said that 'where there is environmental degradation, or acute scarcity of vital resources, war may follow'. Similarly, the then Norwegian Defense Minister Johan Jorgen Holst (1989: 123) argued that environmental stress seems likely to become an increasingly potent contributing factor to major conflicts between nations. In

---

[2]Earlier versions of this article were presented at a NATO Advanced Research Workshop 'Conflict and the Environment' at Bolkesjø, Norway; a meeting of the NATO Committee on the Challenges of the Modern Society Pilot Study Meeting on Environmental Security in Ankara; the Fifth National Conference in Political Science at Geilo, Norway; at the 38th Annual Convention of the International Studies Association, Toronto; and at the 1997 Open Meeting of the Human Dimensions of Global Environmental Change Research Community at Laxenburg, Austria. Financial support from the United States Institute of Peace is gratefully acknowledged, as is assistance from Norunn Grande, Håvard Hegre, Cecilie Sundby, and Bjørn Otto Sverdrup. I am also grateful for comments from guest editor Paul Diehl and two anonymous *JPR* referees, as well as Tanja Ellingsen, Scott Gates, Dan Smith, and a number of participants at the various conferences.

[3]Recent literature surveys are found in Rønnfeldt (1997) and Smith/Østreng (1997).

addition, 'environmental degradation may be viewed as a contribution to armed conflict in the sense of exacerbating conflicts or adding new dimensions'. McMichael (1993: 321) believes that 'the end-stage of unequal power relations and economic exploitation in the world will be tension and struggle over life-sustaining resources. Fossil fuels, freshwater, farming and fish have already become the foci of armed struggles'. Also on an alarmist note, Kaplan (1994), in a widely publicized article in the *Atlantic Monthly*, predicted a coming world anarchy—sparked in large measure by environmental degradation. The Secretary-General of the Habitat conference in 1996 told participants that 'the scarcity of water is replacing oil as a flashpoint for conflict between nations...' (Lonergan 1997: 375).

To this thinking, the prime resource seen as worth fighting for is obviously territory, as in the conflict-filled expansion of European settlers in North America or the border conflicts between China and several neighbors. A more recent variety of territorial conflict concerns the economic zone on the continental shelf, a matter of dispute between most countries which are neighbors at sea and which may raise tiny islands to monumental importance because of their consequences for boundaries at sea. Thus, there are no less than six claimants to all or part of the Spratly Islands in the South China Sea (Denoon/Brams 1997), and the use of force cannot be ruled out. Another is strategic raw materials: the strategic importance accorded to Indochina in the 1950s was justified by US President Eisenhower—in the statement that made famous the 'domino theory'—with reference to the importance of raw materials such as tin, tungsten, and rubber.[4] Some such raw materials are closely tied to arms production, others are simply seen as essential to the economy. A third is sources of energy, the most obvious example being oil supplies from the Persian Gulf, a factor in several recent conflicts. A fourth is water, such as the Atatürk dam project in Turkey, which may result in a water shortage in Syria; or the Nile project, which might provoke a serious downstream-upstream conflict between Egypt and Ethiopia. A UN study identified 214 major river systems shared by two or more countries, many of them subject to unresolved disputes (Renner 1996: 619). A fifth resource is food, including grains and fisheries. Disagreements regarding shared fisheries resources have occasioned numerous confrontations between fishing vessels and armed vessels of coastal states, including three 'Cod Wars' between Iceland and the United Kingdom in the period 1958–76 (Bailey 1997). Increasing food prices have given rise to domestic violent riots; at the international level, food sales have been used for strategic leverage.

The basic causal chain in this argument runs as follows: population growth & high resource consumption per capita $\rightarrow$ deteriorated environmental conditions $\rightarrow$ increasing resource scarcity $\rightarrow$ harsher resource competition $\rightarrow$ greater risk of violence.

---

[4]Statement made by President Eisenhower in a press conference on 7 April 1954, cf. (1954) *Public Papers of the Presidents of the United States. Eisenhower*: 382, quoted from LaFeber (1980: 163). For other statements by US policymakers in the same vein, see Kolko (1985: 76), who finds such references to raw materials to be 'integral to American policy considerations from the inception'.

Not everyone includes all the elements of this causal chain or puts the emphasis in the same place. Biologists frequently single out population growth as the key causal factor; environmentalists tend to start with environmental degradation; and critics of capitalism tend to emphasize excessive consumption in the First World and the need for the First World to restrict its consumption if the Third World is to be allowed to catch up. There is not necessarily any contradiction between these positions, but they stress different parts of the causal chain. Homer-Dixon and associates use a tripartite division of scarcity (Percival/Homer-Dixon 1998: 282–285): supply-induced (which corresponds to environmental deterioration above), demand-induced (resulting from population growth), and structural (due to inequality, which is not included in the model above).

Despite numerous pronouncements on the relationship between conflict and the environment, there is no consensus on the causal mechanisms. Indeed, several writers have questioned the overall argument. Deudney (1991) and Simon (1996) have listed a number of problems with the notion of increasing resource scarcity.[5] First, it ignores human inventiveness and technological change, both of which have vastly increased agricultural yields and the rate of resource extraction from raw material lodes. Modern industry is high on processing, which essentially means intensive in capital, technology, and energy, rather in raw materials like minerals. Second, the pessimistic argument overlooks the role of international trade; most scarcities are local rather than universal. Third, raw materials can be substituted, so being dependent on a particular resource today is not the same as being vulnerable tomorrow if the supply lines should be choked off. Fourth, in the event of increasing scarcity, prices are likely to rise, leading to greater economizing, and further technological change, trade, and substitution. In fact, however, these processes have been sufficiently effective in recent decades for raw materials prices to fall even though global consumption of natural resources has increased. Thus, while the international best-seller *The Limits to Growth* (Meadows et al. 1972) predicted imminent scarcities in a number of minerals, such as copper, the trend since then has in fact been in the opposite direction.

Even in the event of scarcities which could theoretically be overcome by imperialist behavior, the major powers have learnt—from Vietnam, Afghanistan, and a number of other wars in the Third World—that subduing a resisting population, however technologically backward, is a very costly affair. On the basis of an overall 'cornucopian' view where the human being is the most essential resource, Simon (1989) argues that, rather than furthering war, population growth is likely to end it. Instead of armed conflict, Deudney argues, conflicts over resources such as water may lead to joint exploitation of the resources and a network of common interests. Similarly, resource scarcity based on environmental degradation would encourage joint efforts to halt the degradation. Levy (1995) also argues that

---

[5]For general broadsides against environmental pessimism, see Maddox (1972), Bolch/Lyons (1993), Bailey (1995), and Easterbrook (1995). A recent response is found in Ehrlich/Ehrlich (1996).

environmental degradation is unlikely on its own to be a major cause of armed conflict; further, that it is not a national security issue for the United States, and is even unlikely to prove interesting as a research area unless seen in conjunction with other causes of armed conflict.

Many of the more balanced statements on environmental factors in conflict are rather cautious about drawing causal links. Westing (1986: 6), for instance, concludes only that 'what the ultimate cause or causes of war might be defies simple explanation and is, at any rate, far beyond the scope of this analysis'. Brock (1991: 410) concedes that the importance of natural resources as a source of conflict is easily exaggerated, citing Lipschutz/Holdren (1990: 121), who argue that despite Eisenhower's famous 'domino' statement and numerous other policy pronouncements, the problem of access to resources has not 'really played such a central role in shaping US foreign and military policy in recent decades'. The same holds for other nations, Lipschutz/Holdren argue, although resources have frequently served as rationalizations for public consumption 'in support of policies with much more elaborate origins' (Lipschutz/Holdren 1990: 123).

Nevertheless, the overall impression of this literature is one of strong pessimism, stated with considerable force. The object of this article is to examine the research foundations for such claims. I begin with a brief summary of systematic research in this area, and go on to discuss nine common theoretical and methodological problems in the extant literature. Finally, I point to some recent work which seems to be moving in a promising direction and outline some priorities for developing this work further.

6.2   Systematic Research

Neither in the environmental literature nor in studies of the causes of war or civil war has there been much systematic research (quantitative or comparative) on the relationship between resources and environmental factors and armed conflict.

A number of studies summarized in Tir/Diehl (1998) have related population density and population growth to conflict and violence. Strictly speaking, these are not measures of either resource scarcity or environmental degradation. But they may provide a good indirect measure, in that a high value indicates a high or increasing load on resources and the environment. Tir and Diehl found that the literature did suggest a link between population variables and international conflict, but that there was little theoretical or empirical consensus beyond that. In their own empirical study of all nations for the period 1930–89, they conclude that there is a significant but fairly weak relationship between population growth and interstate militarized conflict and war, but that population density has no effect.

Territory is undoubtedly the resource studied most extensively in the context of conflict. Numerous studies, including several quantitative ones, have underlined the role of territorial issues in armed conflict. For example, Holsti (1991: 307) concludes that among interstate wars in the period 1648–1989, territorial issues were by far the most important single issue category: initially figuring in about half the wars, they eventually declined eventually to one-third in the post-World War II period. Only for the period 1815–1914 is territory tied for first place with an issue he calls 'maintain integrity of state/empire', arguably in itself a form of territorial conflict. Reanalyzing Holsti's data, Vasquez (1993: 130; 1995: 284) finds that between 79 and 93 % of wars over the five time periods involve territorially-related issues. Huth (1996: 5) in a study of territorial disputes 1950–90 characterizes this issue as 'one of the enduring features of international politics'. The territorial explanation of war also fits in with the finding that most interstate violence occurs between neighbors (Bremer 1992) or proximate countries (Gleditsch 1995). It is not always obvious why such conflicts occur—is it because neighbors are more easily available for conflict than other states, because of friction in their day-to-day interaction, or because of disputed boundaries or territories? However, Vasquez (1995) presents a strong case for the territorial explanation.

On the other hand, even where territory is conclusively shown to be a significant factor in armed conflict, the question remains whether the territory itself is at issue, or the resources which may be found on it. For the general question which we investigate here, either version will do. But for more precise theorizing about the link between resources and conflict, we need to understand exactly which resource is decisive. Some resources are probably too trivial to fight over, while a resource such as oil might be seen as economically essential. The territory itself might be seen as important to the identity of a people and the symbolic function might be more important than any material value. In a study of modern border disputes, Mandel (1980: 435) hypothesized that ethnically-oriented border disputes would be more severe than resource-oriented ones because ethnic issues seemed less tractable, more emotional, and less conducive to compromise. He was able to confirm his hypothesis in a study of interstate border disputes after World War II, using data from Butterworth (1976). To extend the concept of 'resources' to include ethnic affiliation or the symbolic value of 'the land of our fathers', would be possible, but strained.

A rare empirical investigation of resource imperialism is found in a study by Hammarström (1986, 1997), which examined how interventions in the Third World by three major Western powers (France, UK, and the USA) in the period 1951–77 might be accounted for by the presence of economically and militarily essential minerals in the less developed country. Hammarström's results were essentially negative: the importance of the less developed country as a supplier of minerals did not affect the likelihood of intervention from the USA and the UK, and affected it only slightly in the case of France. This finding also held for the subset of countries within the sphere of influence of the major power, for the subset of minerals upon which the major power was extremely dependent, and for regions rather than individual countries. Hammarström cautions that he has tested the theory on the basis of the theory of economic imperialism only, and that it might also have been analyzed from the perspective of the East-West conflict. But since the major Socialist powers had been largely self-sufficient in raw materials, he felt that such a test would be unlikely to produce very different results.

Anthropologists have studied the influence of environmental factors on tribal warfare in single cases. For instance, Graham (1975) attributes considerable importance to environmental factors in the explanation of the endemic intertribal warfare among the Yuman societies of the Colorado and Gila rivers. As far as modern warfare is concerned, however, there appears to be little systematic evidence. For example, the comprehensive *Handbook of War Studies* (Midlarsky 1989) does not list 'ecology', 'environment', 'land', 'raw materials', or 'water' in its index.[6] Neither do the indices of such classical studies of war as Richardson (1960) or Wright (1942/1965).[7] Within the Correlates of War project—the largest modern research project of its kind—one article finds limited support for the idea that population pressure may be a factor in war initiation (Bremer et al. 1973); but generally environmental factors do not seem to have attracted much attention. Choucri/North (1975) have also investigated the effects of population growth in the international processes that led up to World War I. However, in a more recent wide-ranging book by North (1990), both the environment and war are discussed extensively but the two seem hardly to intersect. In general, those who have researched the general patterns of war have been much more concerned with alliances, power configurations, and other elements of realist theory (and more recently with democracy, economic interdependence, and other elements of liberal theory) than with environmental factors. It is possible, of course, that this is because environmental factors simply do not play much of a role in warfare—but one would feel more confident of this conclusion if environmental hypotheses had at least been tested. Another explanation for the relative neglect of these factors could be that the environmental boundaries of state policy have not been central to the grand political debate until quite recently. Moreover, most research on international conflict has focused on national, dyadic, and systematic attributes for understanding international behavior, whereas the issues involved in conflict have generally been ignored —including, presumably, environmental issues (Diehl 1992).

Domestic armed conflict is dominant in the single case studies on the effects of environmental degradation. But there is even less comparative and quantitative work here than in relation to interstate conflict. Wallensteen/Sollenberg (1997: 343), in a study of armed conflicts after the Cold War (the vast majority of which have been domestic) show that in slightly more than half the conflicts the basic incompatibility concerned territory rather than government. Conflicts over territory were less frequently terminated (or only tentatively terminated with a ceasefire) and were less frequently the subject of peace agreements. The article by Hauge/ Ellingsen (1998) stands out as fairly unique in trying to integrate environmental

---

[6]More recently, Midlarsky (1995) has investigated how the lack of warfare and two environmental variables (rainfall and sea borders) exert positive effects on democracy and the impact of democracy on environmental policies (1998).

[7]With the exception that Wright's book contains a few references to environmental factors in 'primitive warfare', for example, that 'primitive peoples in extremely cold and extremely hot climates tend to 'be unwarlike', while in general 'a temperate or warm, somewhat variable and stimulating climate favors warlikencss' (Wright 1965: 63, 552–554).

degradation into a more general model of civil war and test it in a large-N mode. They conclude that environmental degradation does stimulate the incidence of conflict, but less so than political or economic variables and that the severity of such conflicts is better accounted for by military spending. Their study is limited to three types of environmental degradation which mainly affect poor countries and covers a relatively limited time-span (1990–92).

6.3 Nine Common Problems

Apart from the role of population factors and territory in armed conflict there is, then, a notable lack of systematic research on the effects or resource or environmental factors on armed conflict. In the absence of solid evidence, the field has been left wide open to speculation and conjecture. Thus, in debating population pressure, even serious academics are driven to support their respective positions in the US debate by referring to the sparsity of population that anyone can observe out of an airplane window (Bolch/Lyons 1993: 27) or the obvious overpopulation which is evident when one drives in a major city at rush hour (Ehrlich/Ehrlich 1996: 211). Such low standards of evidence make it difficult to assess the state of the art. In what follows, I will concentrate on work with more solid claims to seriousness. Even within this literature, however, there are many problems. This article discusses nine of them, in no particular order.

6.3.1 Resource Scarcity or Environmental Degradation?

Many of the references to 'environmental' factors that are posited as capable of stimulating an arms race or triggering a war are unclear as to whether the causal factor is absolute resource scarcity or environmental degradation. Virtually all resources are 'scarce'—to some degree, at some times, or in some places. By definition, scarcity leads to conflict in the sense of conflict of interest. It can even be argued that all conflicts of interest derive from scarcity. However, not all resource conflicts lead to overt conflict behavior, and even fewer to the use of force. Environmental degradation may exacerbate resource conflicts because it reduces the quantity or quality or the resource in question. Pollution of a river, for instance, reduces the quality of the water; but it can also be interpreted as reducing the quantity of clean water, and therefore contributing to increased scarcity. Similarly, air pollution in a city degrades the quality of the air and changes an unlimited public good (clean air) into a scarce one.

Libiszewski (1992: 2) argues that simple resource conflicts are very common, but that the concept of environmental conflict calls for a more restricted use. The latter he defines as a conflict caused by a human-made disturbance of the normal regeneration rate of a renewable resource (Libiszewski 1992: 6). Thus, a conflict over agricultural land is an 'environmental conflict' if the land becomes an object of contention as a result of soil erosion, climate change, and so on, but not in the case of an ordinary territorial or colonial conflict or an anti-regime civil war aiming at the redistribution of land. Non-renewable natural resources (such as oil) are not integrated in any eco-system. Their depletion may lead to economic problems, but they are not in themselves environmental problems, so conflicts over such resources should not be considered environmental conflicts.

Libiszewski's distinction between those conflicts which result from simple resource scarcity and those which result from environmental degradation is useful. When, for instance, Homer-Dixon (1991, 1994) refers to 'environmental scarcity', the terminology itself muddies the waters. In the following, within the bounds of the practical, I will try to keep the two apart. However, I find it more difficult to follow Libiszewski in linking environmental conflict to the concept of an eco-system, with its questionable overtones of balance and equilibrium.[8]

Even the distinction between simple resource scarcity and environmental degradation raises some problems. Today's simple scarcity may well be the result of environmental mismanagement in the past. The lack of forests around Madrid may be seen as a fact of nature today, but can be interpreted as a result of excessive ship-building in the 16th century, and thus as an old case of environmental degradation. Most, if not all, territorial conflict can be seen as the result of past population policies (or a lack thereof) which have permitted groups to multiply beyond what their traditional territories could support. As far as the present is concerned, however, this distinction sets a useful standard.

6.3.2 Definitions and Polemics

The term environmental security was launched to place the environment on the agenda of 'high politics' (Lodgaard 1992: 113). If one adopts a broad conception of security as 'the assurance people have that they will continue to enjoy those things that are most important to their survival and well-being' (Soroos 1997a: 236), it can be plausibly argued that serious environmental degradation can indeed threaten security. This would be particularly true if the most serious warnings about global warming or holes in the ozone layer turn out to be correct, but even more traditional environmental concerns like air and water pollution can kill more people than smaller armed conflicts or even wars. Politically, then, it makes sense to give such

---

[8]Libiszewski speaks (1992: 3) of 'a dynamic equilibrium oscillating around an ideal average'. Whether such equilibria exist in anything but the short term, seems questionable. At least it is implausible that only human intervention can change them. Otherwise, it would be difficult to explain the disappearance of the dinosaurs and other animals long before human beings were numerous enough to have much influence on the global environment, or even before human beings existed. Or should we see the emergence of the dinosaurs, as well as their subsequent disappearance, as part of an 'ideal' world?

issues very high priority. Like common security, structural violence, or sustainable development, environmental security made a good slogan—so successful that the US Department of Defense now has a position called 'Principal Deputy Undersecretary of Defense (Environmental Security)', the NATO Science Committee is running a series of workshops on environmental security (Gleditsch 1997), and NATO's Committee on the Challenges of Modern Society is conducting a pilot study on the same topic (Carius et al. 1996). Defense establishments in many countries in NATO and among the cooperating partners in East and Central Europe are rushing to acquire a green image by improving their environmental performance.

But political success does not necessarily make a slogan into a workable research tool (Graeger 1996). Merging two objectives like security and environmental protection into a joint term does not give us new theoretical or empirical insight into whether the two are mutually supportive—or in competition. Those who on the basis of the broad definition of security deliberately disregard the question of armed conflict are in a sense on fairly safe ground.[9] But most of the literature cannot resist the temptation to bring the danger of armed conflict back in, as a consequence of environmental degradation (Gleditsch 1997; Lodgaard 1992). Indeed, why else would armed forces and military alliances be so interested in environmental security?

On this point, the critical literature (Deudney 1990; Levy 1993) does not take us much further. In part, this literature engages in similar definitional exercises in order to prove the futility of the concept of environmental security. In addition, it demonstrates theoretical and empirical problems in the writings on the environment and security. Some of the critical points are well taken, but if they do not end up in an alternative or improved research design, they are of little help.

6.3.3 Overlooking Important Variables

If we could prove that human activity could shift the average global temperature by, for example, 5°, this would be a very important finding. No climate researcher would argue, however, that human activity is the one and only determinant of global temperature. Anyone who correlated emissions of greenhouse gases with temperatures recorded monthly would seem patently ridiculous, since the effect of human activity is likely to be completely swamped by long-established seasonal variations. In the social sciences, such caution is often thrown to the winds. Far too many analyses of conflict and the environment are based not only on bivariate analysis but also on overly simplistic reasoning.

---

[9]There is still a danger of conceptual slippage, by including all manner of environmental problems in the concept of environmental security, and not just those which are serious enough to be treated on a par with war destruction. This development is reminiscent of the fate of the concept of structural violence (Galtung 1969), which was so successful in the short term that it came to include any social ill—and eventually self-destructed.

The greatest weakness in this respect is that much of this literature ignores political, economic, and cultural variables. When writers on environmental conflict refer to the '214 shared river systems' as potential sources of conflict, they rarely distinguish between rivers which run through poor, undemocratic, politically unstable countries ridden by ethnic tensions, and rivers running through stable and affluent countries. It is tacitly assumed that resource conflicts have a high potential for violence, regardless of the countries' political system or economic orientation. Since democracies rarely if ever fight one another (Gleditsch/Hegre 1997; Russett 1993) and since they rarely experience civil war (Hegre et al. 2001) there is no reason to believe that they will suddenly start fighting over resource issues between themselves, or internally, any more than over other issues. Moreover, if it is correct that democracies generally display more benign environmental behavior than do non-democracies,[10] then democracies are also less likely to generate the kind of extreme environmental degradation which may be assumed to generate violent conflict. Thus, democracy may have a double effect in preventing armed conflict over the environment: it generates fewer serious problems, and it provides other means of conflict resolution once these problems have arisen.

Most work on environmental conflict does not discuss how regime type may influence such conflict. For example, in the many case studies published as project reports by Homer-Dixon and his associates, there are general references to 'key social and political factors' (Percival/Homer-Dixon 1993: 3), to corruption, weakened legitimacy, resource capture by elites, and so on. However, words such as 'democracy' or 'autocracy' do not occur in the model. In view of the extensive theoretical and empirical literature relating the degree of democracy to civil violence in an inverted U-curve (Muller/Weede 1990) a democracy variable should have been included explicitly. The reports frequently hover around the idea that democratic procedures might have something to do with the level of conflict. Yet, none of the reports clearly state that democracy matters, or in what way. Furthermore, the work by Homer-Dixon and his associates is on the whole *more* sensitive to political variables than most studies in this field.

Many of the militarized interstate disputes between democracies have been over resources—or more specifically over one particular resource, fish (Bailey 1997; Russett 1993: 21) At sea, boundaries between states are not yet well settled, and even where they are established by law or by custom, they are not visible. The fluidity of any sea boundary makes it more conflict-prone than an established land boundary. Moreover, fish stocks straddle national boundaries and migrate across them with the seasons, with no concern for the consequences for human conflict. It is not surprising, then, that international fisheries should be ridden with conflict. However, even if fisheries conflicts between democracies may involve some use of force or threats to use it, such conflicts rarely, if ever, escalate to the point where human life is lost. Since 'war' is usually defined as a conflict with more than 1,000

---

[10]As argued by Payne (1995), Gleditsch/Sverdrup (1995), and Gleditsch (1997a); Midlarsky (1998) is more sceptical.

dead (Small/Singer 1982; Wallensteen/Sollenberg 1997), terms such as 'cod war' or 'turbot war' (Soroos 1997b) are misnomers. Moreover, these conflicts usually involve one private party (a fishing vessel) against representatives of another state (a warship or a coastguard vessel). When such conflicts occur between democracies, the two states take particular care not to engage in force or threats of force between their own representatives. Thus, as far as the militarized part of the conflicts is concerned, these disputes are not really 'interstate'.

A similar point holds for economic variables. Much of the environmental literature lacks explicit recognition of the fact that material deprivation is one of the strongest predictors of civil war. Moreover, economically highly developed countries rarely fight one another (Mueller 1989), although this regularity is less absolute than the democratic peace. Finally, while economic development does tend to exacerbate certain environmental problems (such as pollution and excessive resource extraction) up to a point, the most advanced industrial economies also tend to be relatively more resource-friendly. Hence, resource competition is likely to be less fierce domestically as well as externally among the most highly developed countries. Going back to the example of shared water resources, highly developed countries have very strong economic motives for *not* fighting over scarce water resources; instead, they use technology to expand the resources or find cooperative solutions in exploiting them. Poor countries generate more local environmental problems, which in turn may exacerbate their poverty and which is also conducive to conflict. Certain types of environmental degradation—like deforestation, lack of water and sanitation, and soil erosion—are part and parcel of underdevelopment.

6.3.4 Untestable Models

While there is much single-factor reasoning, some work goes to the opposite extreme. In a series of reports and articles which represent some of the most solid case-oriented work in the field, Homer-Dixon (1991, 1994) employs a very complex theoretical scheme, where four basic social effects of environmental disruption (decreased regional agricultural production, population displacement, decreased economic productivity, and disruption of institutions) may produce scarcity conflicts, group-identity conflicts, and relative-deprivation conflicts. This model has been reproduced in various forms in a number of publications from the Toronto Group, and by others (cf. Hauge/Ellingsen 1998: 301).

Some problematic aspects of these complex models are clearly seen in the case studies from Homer-Dixon and his associates. The rebellion in Chiapas (Howard/Homer-Dixon 1995), for example, is explained by seven (mostly economic) independent variables acting through nine intervening variables and one additional independent variable. Violence in Gaza (Kelly/Homer-Dixon 1995) involves an explanatory scheme of eight independent and intervening variables, which in turn draw on a six-variable scheme for explaining three kinds of water scarcity and a ten-variable scheme for explaining the increasing level of grievance against the Palestine National Authority. Whether in a large-N or a comparative case study mode, such a comprehensive scheme would be very difficult to test. Empirical testing is not helped by the fact that many of the variables are rather imprecise, such as 'health problems'. Similar problems can be found in the work of Lee (1996) who has done interesting case studies of Sudan and Bangladesh.

Of course, single-factor reasoning and overly complex models do not go together. But the joint effect of the two phenomena is a lack of modestly multivariate analyses and of a gradualistic approach to model-building. This is not an argument against the development of large and complex models, like those developed in macroeconomics, some of which have also been applied to environmental problems (Nordhaus 1994). But such models must be built gradually, with more limited modules being put to the test first.

6.3.5 The Lack of a Control Group

Qualitative and quantitative research serve the same logic of inference, although their styles are different (King et al. 1994: 3). In the literature on the environment and armed conflict, the case study has been by far the dominant approach. Homer-Dixon (1991: 83) criticized earlier writing on this topic as 'anecdotal' and has added (Homer-Dixon 1994) a number of carefully documented case studies analyzed on the basis of his detailed theoretical scheme. Levy (1995: 56) argues that Homer-Dixon's case studies offer 'more anecdotes, but not more understanding'. Recent studies from Homer-Dixon's project deal with Gaza (Kelly/Homer-Dixon 1995), South Africa (Percival/Homer-Dixon 1995, 1998) and Chiapas (Howard/Homer-Dixon 1995). Similarly, the Environment and Conflicts Project at the Swiss Federal Institute of Technology has carried out a number of case studies, recently published in three volumes (Bächler et al. 1996).

The charge that such case studies are anecdotal cannot easily be dismissed, in that all of them are single case studies of 'environmental scarcity and violent conflict'. They offer no variation on the dependent variable,[11] thus violating an important principle of research design, whether the approach is qualitative or quantitative (King et al. 1994: 108). Other projects based on single case studies (e.g. Lee 1996) suffer from the same problem. Regardless of the accuracy of the historical description and the excellence of the theoretical model, these studies fail to provide an empirical basis for comparison. In the Toronto Group's study of Chiapas, for example, 'weak property rights' is a factor in creating 'persistent structural scarcities' which in turn contributes to the outbreak of rebellion (Howard/Homer-Dixon 1995: 23). But in order to evaluate the causal nature of this link, we need to examine cases without conflict, many of which will certainly also

---

[11]Nor, for that matter, do they provide any variation on the independent variable, but that is not necessarily a problem in the research design, cf. King et al. (1994: 137).

be characterized by weak property rights. Only when we know that conflict occurs more frequently in the former group, can we even begin to think about causal links.

Homer-Dixon and associates justify this method by arguing that 'biased case selection enhances understanding of the complex relationships among variables in highly interactive social, political, economic, and environmental systems' (Percival/Homer-Dixon 1998: 279; Homer-Dixon 1996). There are two problems with this argument. One is that it seems to imply that environmental problems are more complex than other social (or for that matter physical) phenomena that researchers study. No justification is given for this view. I would argue, on the contrary, that any social system is as complex as the theory developed to study it. In other words, the complexity is in the mind of the beholder rather than in the phenomenon itself. Second, almost any methodological limitation can be justified at an exploratory stage. The problem arises if the project does not move on to the next stage, but instead concludes on the basis of the exploratory case studies that 'environmental scarcity causes violent conflict' (Homer-Dixon 1994: 39).

Even some of the best quantitative studies on resources and war, those on territorial conflict, suffer from the same problem. Both Holsti (1991) and Vasquez (1993) derive their findings concerning the territorial basis of armed conflict from an examination of the issues involved in the armed conflict. However, they do not examine situations which did not escalate to armed conflict to see if also they contained unresolved territorial issues. Huth (1996), who studies territorial disputes and not just wars, does not include territorial claims which are not expressed publicly (Huth 1996: 24, 239). For example, the question of the Finnish territories conquered by the Soviet Union in the Winter War of 1939–40 could not be raised during the Cold War due to Finland's somewhat precarious position. Thus, if one wanted to test a hypothesis about conditions under which territorial disagreements are completely suppressed, Huth's dataset would not be suitable.[12]

In examining only cases of conflict, one is likely to find at least partial confirmation of whatever one is looking for, unless there are very clearly specified criteria for the threshold level of the independent variable assumed to lead to violence. No society is completely free of environmental degradation, nor is any society completely free of ethnic fragmentation, religious differences, economic inequalities, or problems of governance. From a set of armed conflicts, one may variously conclude that they are all environmental conflicts, ethnic conflicts, clashes of civilizations, or products of bad governance. Indeed, conflicts like the internationalized civil war in Ethiopia from the mid-1970s have been described in most of these terms. Only by adopting a research design where cases of conflict and non-conflict are contrasted can the influence of the various factors be sorted out.

---

[12]Another problem, peculiar to the literature on territory and armed conflict, is that regardless of the issue which started the conflict, the contestants need a territorial base to deploy force of any size; even guerrillas need some sort of safe haven. Thus, armed conflicts, domestic as well as international, at least when they escalate to a certain size, *become* conflicts over territory even if territory was not the most salient issue from the start.

6.3.6 Reverse Causality

It is well established—and in a sense not very surprising—that modern war wreaks havoc on the environment (see e.g. Westing 1990, 1997). The Vietnam War brought this issue to public attention, although earlier large wars had also caused destruction of vital infra-structure and generated other negative environmental effects. More recently, the prospect of a 'nuclear winter' pointed to the prospect of the obliteration of human activity on the Northern hemisphere as a result of the environmental effects of a nuclear war. For instance, Sagan/Turco (1993) maintained that a global nuclear war could lead to a worldwide fall in temperature of 15–20 °C. Even more optimistic scenarios than this could put the earth's climate back to the most recent ice age. These environmental effects could be worse than the direct impact of nuclear war such as blast, fire, and radioactive fallout. Today, the campaign to abolish landmines has focused public attention on the long-term environmental effects of a weapon long after its military utility has gone.

This war-environment relationship is sometimes confused with the possibility that environmental degradation *causes* armed conflict and war. For instance, in arguing for a link from the environment to violent conflict, Holst (1989: 126) points out that five of the six countries on the UN list of countries most seriously affected by hunger had experienced civil war (Ethiopia, Sudan, Chad, Mozambique, and Angola). However, it is highly probable that the violent uprising contributed to the hunger, or even that starvation was used as a weapon of war in some of these conflicts. Thus the most important causal link is very likely the opposite of that indicated by Holst.

A slightly more complicated relationship is suggested by McMichael (1993: 322) as a positive feedback process: 'environmental destruction and resource scarcity promote war which, when it breaks out, further increases environmental destruction and resource depletion'. However, a somewhat different feedback process seems more likely:

> war → environmental destruction → resource conflict
> → exacerbated armed conflict

This process starts from a well-documented relationship rather than from a more conjectural one. It also contains in the endpoints the process of violence repeating itself over time, which has been found to be highly significant in studies of interstate war (Raknerud/Hegre 1997) as well as civil war (Hegre et al. 2001). Of course, if the process were to continue indefinitely, these two feedback cycles would be identical. Moreover, the world would have entered a process of accelerating deterioration and violence. Studies of interstate war and civil war indicate that violence is repeated, but not always, and not as a rule with increased intensity. Rather, it may be thought of as an echo, always weaker than the signal it reflects, and petering out in the end. Thus, it does matter whether the process starts with war or with environmental degradation.

6.3.7 Using the Future as Evidence

Homer-Dixon, and many other authors in this area, have stressed the potential for
violent conflict in the future. There is a lack of empirical study of armed conflicts in
the past as well as a lack of explicit theorizing for if and why resource scarcities
should have a higher violence-generating potential in the future than in the past.
Much of the literature deals with conflicts of interest involving potential violence
rather than with actual violence. For example, no one is really arguing that any
armed conflict in the past has occurred mainly because of water issues. To argue
that water has been a main issue in the many conflicts in the Middle East, and
specifically in the wars between Israel and its neighbors, would be to seriously
underestimate the explosive ethnic and territorial issues in the region (Lonergan
1997: 383). The argument is entirely in terms of future wars which may happen. In
Silent Spring, arguably the most influential environmentalist book ever published,
Carson (1962: Chap. 1) described in the past tense 'a town in the heart of America'
hit by mysterious diseases caused by the excessive use of pesticides, but in fact this
was 'a fable for tomorrow'. Similarly, when Ehrlich/Ehrlich (1968: 11) started *The
Population Bomb* with a statement that 'The battle to feed all of humanity is over',
went on to predict that hundreds of millions of people would starve to death, and
then discussed the political consequences, they were arguing from future empirical
'evidence' which in fact turned out to be wrong. While they now hold that the
principal problem 'of course' is not acute famine, but malnutrition (Ehrlich/Ehrlich
1996: 76), they also argue that there is every reason to think that the limits to the
expansion of plant yields is not far off (Ehrlich/Ehrlich 1996: 80) and liken the
human race to animal populations which grow beyond their carrying capacity until
they 'crash' to a far lower size (Ehrlich/Ehrlich 1968: 67). These are hypotheses
based on controversial theory and debatable extrapolations, rather than 'data' which
may confirm the predictions.

 In principle, the future may always differ from the past. Despite whatever
painstaking empirical mapping we may have made of past wars, future wars may
run a different course. Environmental organizations and other advocacy movements
are prone to argue that we are now at a turning-point in human history and that
patterns from the past may no longer hold in the future. In saying this, one may
easily slip into prophecy. 'There will be water wars in the future' is no more a
testable statement than the proverbial 'The End of the World is at Hand!', unless
terms such as 'the future' and 'at hand' are clearly specified. In an effort to make
pessimistic environmental predictions more precise (and to prove them wrong) the
economist Julian Simon has repeatedly challenged his opponents to place bets on
resource issues. Three environmentalists took him up on this in 1980 and bet that
the price of a basket of five metals would rise over a ten-year period. Simon, who
thought they would decline, ended up winning the bet (Myers/Simon 1994: 99,
206). To my knowledge, no one has issued bets on environmental degradation and
warfare, but conceivably this might be a useful strategy in order to provoke greater
scholarly precision.

6.3.8 Foreign and Domestic Conflict

Since the end of World War II a large majority of wars have been domestic rather than interstate (Gleditsch 1996: 294). Although the number of domestic armed conflicts, whether the smaller ones or the larger conflicts conventionally called 'wars', has declined slightly after the end of the Cold War (Wallensteen/Sollenberg 1997), they remain much more numerous than international armed conflicts. This pattern is unlikely to be broken in the foreseeable future.

Homer-Dixon's work is explicitly related to domestic conflicts, and Tir/Diehl (1998: 319) argue that most studies of population pressures and war focus on internal conflict. Yet, much of the reasoning about the prevalence of scarce resources as a factor in war is built on lessons from the study of interstate war, as my literature review above indicates. Both at the theoretical and empirical level, the study of interstate conflict has been conducted largely separately from the study of civil war. Some factors are similar, but one cannot easily generalize from one to the other. An obvious difference is that many theories of war at the interstate level are related to the absence of any overarching system of power, i.e. what realists call international anarchy. At the domestic level, war is often related to revolt against excessive state power or its abuse. Many issues which stimulate armed conflicts at the interstate level may be too weak to force a break *within* a society held together by a central authority. Theories linking environmental degradation to violence therefore need to be quite specific concerning whether they are addressing domestic or interstate violence. At this stage it is probably appropriate to have separate explanatory models for the two phenomena—at least in the absence of some bold theoretical thinking concerning how to link theories of violence at the domestic and interstate levels.

6.3.9 Levels of Analysis

Studies of war require precision about the unit of analysis. For example, in studies of democratic peace, it is frequently assumed that if democracies do not fight one another, then there will be more peace as the fraction of democracies grows. I have shown elsewhere (Gleditsch/Hegre 1997), that this holds true only under certain conditions. Under a plausible set of assumptions, an increase in the number of democracies is more likely to lead initially to an increase in the frequency of war in the system. Only later, after the degree of democracy is above a certain level, will further democratization decrease the probability of war. Similarly, we cannot automatically generalize theories and empirical evidence concerning resource and environmental factors from one level to another. Below, follow three hypotheses about interstate armed conflict using the same independent and dependent variables, but at different levels of analysis:

(1) *System level*: In a world with high resource constraints, there will be more interstate conflict.
(2) *Nation level*: Countries with high resource constraints are more likely to be involved in conflicts with other countries.
(3) *Dyadic level*: Countries with high resource constraints are likely to be involved in conflict with countries with an ample supply of the same resource, and even (but to a lesser extent) with other countries with the same resource constraints.

Although these three hypotheses are derived from the same kind of thinking, the one does not logically follow from the other. If we assume that the overall frequency of interstate war is regulated mostly by systemic factors (such as the balance of power), or that states' propensity to war is largely determined by national characteristics like regime type or wealth, resource factors may still determine the direction of warfare (i.e. dyadic war). Thus, even if resource scarcities are relevant for 'who fights whom', that is not equivalent to saying that global resource scarcity determines the overall level of armed conflict. This problem of levels is not, to my knowledge, dealt with at any length in the relevant literature, which freely jumps between the dyad, the nation, and the system levels for theory as well as empirical evidence.

6.4 The Way Ahead

The nine problems discussed above add up to a fairly pessimistic assessment of the state of the study of environmental causes of conflict. Even leading studies in the field come up against fairly elementary problems in theory construction or empirical testing. Critical studies, like those of Deudney and Levy, are valuable in pointing out some of these problems. But the critique will serve to advance the field only if it stimulates more satisfactory research.

Systematic cross-national study by social scientists of any aspect of the environment is in its infancy. On the positive side, we may note that economists have done a great deal of research on how economic development drives environmental stress. A common finding in this literature is that the rate of emissions of environmentally harmful products increases with growing wealth, but not linearly; rather the environmental damage tapers off at high levels of development. It is clear that for some noxious emissions (such as SO_2) there is a significant decrease at very high levels of economic development, because rich countries can afford to acquire modern technology and also because their decision-makers put a higher premium on a clean environment.[13]

In recent work on democracy and the environment (Gleditsch/Sverdrup 2002; Midlarsky 1998) attempts have been made to relate indicators of the environmental

---

[13]Dietz/Rose (1997) provide a brief survey of recent writings.

performance of nations to their regime characteristics. The conclusions from these two empirical analysis are at some variance with one another. Generally, the study of political predictors of environmental degradation lags far behind the study of economic factors.

There is even less rigorous work using environmental degradation as a predictor to conflict. The work by Tir/Diehl (1998) and Hauge/Ellingsen (1998) is relevant and representative of a tradition in theoretically-grounded empirical research on armed conflict, based on cross-national (and, to a more limited degree, cross-temporal) data for all nations. Both these analyses place the analysis of resource and environmental variables squarely within a multivariate perspective. Both studies do indeed find an effect of such variables—population growth in one study; deforestation, land degradation, and low availability of fresh water in the other. Since all these predictor variables are traditionally associated with poverty, this raises the issue if the association between conflict and environmental load (as in the Tir and Diehl study) or conflict and environmental degradation (in Hauge and Ellingsens's work) may be primarily an underdevelopment problem. Highly developed (or even 'overdeveloped') countries also have environmental problems (traffic noise, industrial pollution, etc.) but there is no evidence that such environmental issues generate armed conflict, internally or externally. In this sense, perhaps environmental conflict should be analyzed as a development issue? At least this is an avenue worth further exploration.

A striking feature of the existing empirical studies is the problem of gathering valid and reliable data on the environmental behavior of nations or smaller geographical units. Environmental accounting is miles behind national economic accounting. The environmental variables used in the Tir and Diehl and Hauge and Ellingsen studies, and in Midlarsky (1998), are not very highly correlated overall. Is this mainly caused by low data reliability, or because they tap different dimensions of what might be called environmental performance? In order to answer this question, and to move forward in relating environmental studies and the study of armed conflict, we need major improvements in systematic data collection—a Correlates of War project for the environment.

References

Bächler, Günther; Böge, Volker; Klötzli, Stefan; Libiszewski, Stephan; Spillmann, Kurt R., 1996: *Kriegsursache Umweltzerstörung. Ökologische Konflikte in der Dritten Welt und Wege ihrer friedlichen Bearbeitung* [Environmental Destruction as a Cause of War. Ecological Conflicts in the Third World and Peaceful Ways of Resolving Them]. Three volumes (Zurich: Rüegger).
Bailey, Jennifer, 1997: "States, Stocks, and Sovereignty: High Seas Fishing and the Expansion of State Sovereignty", in: Gleditsch, Nils Petter (Ed.): *Conflict and the Environment* (Dordrecht: Kluwer): 215–234.

Bailey, Ronald, 1993: *Eco-Scam. The False Prophets of Ecological Apocalypse* (New York: St Martin's).

Bolch, Ben; Lyons, Harold, 1993: *Apocalypse Not: Science, Economics, and Environmentalism* (Washington, DC: Cato Institute).

Bremer, Stuart, 1992: "Dangerous Dyads. Conditions Affecting the Likelihood of Interstate War, 1816–1965", in: *Journal of Conflict Resolution*, 36,2: 309–341.

Bremer, Stuart; Singer, J. David; Luterbacher, Urs, 1973: "The Population Density and War Proneness of European Nations, 1816–1965", in: *Comparative Political Studies*, 6,3: 329–348. [Reprinted in Singer, J. David et al. (Eds.): *Explaining War. Selected Papers from the Correlates of War Project* (Beverly Hills, CA & London: SAGE): 89–207.]

Brock, Lothar, 1991: "Peace Through Parks: The Environment on the Peace Research Agenda", in: *Journal of Peace Research*, 28,4: 407–423.

Brundtland, Gro Harlem et al., 1987: *Our Common Future. World Commission on Environment and Development* (Oxford: Oxford University Press).

Butterworth, Robert L., 1976: *Managing Interstate Conflict, 1945–1974* (Pittsburgh, PA: University Center for International Studies).

Carius, Alexander; Oberthür, Sebastian; Kemper, Melanie; Sprinz, Detlef, 1996: *NATO CCMS Pilot Study. Environment and Security in an International Context. State of the Art and Perspectives. Interim Report* (Berlin: Ecologic & Potsdam: Potsdam Institute for Climate Impact Research, for German Federal Ministry of Environment, Nature Conservation and Nuclear Safety).

Carson, Rachel, 1962: *Silent Spring* (Boston, MA: Houghton Mifflin).

Choucri, Nazli; North, Robert C., 1975: *Nations in Conflict* (San Francisco, CA: Freeman).

Colinvaux, Paul A., 1980: *The Fates of Nations: A Biological Theory of History* (New York: Simon & Schuster).

Denoon, David B.H.; Brams, Steven J., 1997: "Fair Division: A New Approach to the Spratly Islands Controversy", in: *International Negotiation*, 2,2: 303–329.

Deudney, Daniel, 1990: "The Case Against Linking Environmental Degradation and National Security", in: *Millennium*, 19,3: 461–476.

Diehl, Paul F., 1992: "What Are They Fighting For? The Importance of Issues in International Conflict Research", in: *Journal of Peace Research*, 29,3: 333–344.

Dietz, Thomas; Rosa, Eugene A., 1997: "Environmental Impacts of Population and Consumption", in: Stern, Paul C.; Dietz, Thomas; Ruttan, Vernon W.; Socolow, Robert H.; Seeney, James L. (Eds.): *Environmentally Significant Consumption. Research Directions* (Washington, DC: National Academy Press): 92–99.

Easterbrook, Gregg, 1995: *A Moment on the Earth. The Coming Age of Environmental Optimism* (New York: Viking Penguin).

Ehrlich, Paul R.; Ehrlich, Anne H., 1972: *Population, Resources, Environment. Issues in Human Ecology*, 2nd edn. (San Francisco, CA: Freeman). First published in 1970.

Ehrlich, Paul R.; Ehrlich, Anne H., 1996: *Betrayal of Science and Reason. How Anti-Environmental Rhetoric Threatens Our Future* (Washington, DC & Covelo, CA: Island Press).

Galtung, Johan, 1969: "Violence, Peace, and Peace Research", in: *Journal of Peace Research*, 6,3: 167–191.

Galtung, Johan, 1982: *Environment, Development and Military Activity. Towards Alternative Security Doctrines* (Oslo: Norwegian University Press).

Gleditsch, Nils Petter, 1994: "Conversion and the Environment", in: Käkönen, Jyrki (Ed.): *Green Security or Militarized Environment?* (Aldershot & Brookfield, VT: Dartmouth): 131–154.

Gleditsch, Nils Petter, 1995: "Geography, Democracy, and Peace", in: *International Interactions*, 20,4: 297–323.

Gleditsch, Nils Petter, 1996: "Det nye sikkerhetsbildet: Mot en demokratisk og fredelig verden? [The New Security Environment: Towards a Democratic and Peaceful World?]", in: *Internasjonal Politikk*, 54,3: 291–310.

Gleditsch, Nils Petter, 1997a: "Environmental Conflict and the Democratic Peace", in: Gleditsch, Nils Petter (Ed.): *Conflict and the Environment* (Dordrecht: Kluwer): 91–106.

Gleditsch, Nils Petter, (Ed.), 1997b: *Conflict and the Environment* (Dordrecht: Kluwer).

Gleditsch, Nils Petter; Hegre, Håvard, 1997: "Peace and Democracy: Three Levels of Analysis", in: *Journal of Conflict Resolution*, 41,2: 283–310.

Gleditsch, Nils Petter; Sverdrup, Bjørn Otto, 2003: "Democracy and the Environment", in: Page, Edward A.; Redclift, Michael (Eds.): *Human Security and the Environment: International Comparisons* (Cheltenham: Edward Elgar): 45–70.

Gleick, Peter H. (Ed.), 1993: *Water in Crisis: A Guide to the World's Freshwater Resources* (New York & Oxford: Oxford University Press).

Goertz, Gary; Diehl, Paul F., 1992: *Territorial Changes and International Conflict* (London & New York: Routledge).

Graham, Edward E., 1975: "Yuman Warfare: An Analysis of Ecological Factors from Ethnohistorical Sources", in: Nettleship, Martin A.; Dalegivens, R.; Nettleship, Anderson (Eds.): *War, Its Causes and Correlates* (The Hague & Paris: Mouton): 451–462.

Granger, Nina, 1996: "Environmental Security?", in: *Journal of Peace Research*, 33,1: 109–116.

Hammarström, Mats, 1986: *Securing Resources by Force. The Need for Raw Materials and Military Intervention by Major Powers in Less Developed Countries*. Report no. 27 (Uppsala: Department of Peace and Conflict Research, Uppsala University).

Hammarström, Mats, 1997: "Mineral Conflict and Mineral Supplies: Results Relevant to Wider Resource Issues", in: Gleditsch, Nils Petter (Ed.): *Conflict and the Environment* (Dordrecht: Kluwer): 127–136.

Hauge, Wenche; Ellingsen, Tanja, 1998: "Beyond Environmental Security: Causal Pathways to Conflict", in: *Journal of Peace Research,* 33,3: 299–317.

Hegre, Håvard; Ellingsen, Tanja; Gates, Scott; Gleditsch, Nils Petter, 2001: "Toward a Democratic Civil Peace? Democracy, Political Change, and Civil War, 1816–1992", in: *American Political Science Review,* 95,1: 17–33.

Holst, Johan Jørgen, 1989: "Security and the Environment: A Preliminary Exploration", in: *Bulletin of Peace Proposals,* 20,2: 123–128.

Holsti, Kalevi J., 1991: *Peace and War. Armed Conflicts and International Order 1648–1989* (Cambridge: Cambridge University Press).

Homer-Dixon, Thomas, 1991: "On the Threshold: Environmental Changes as Causes of Acute Conflict", in: *International Security,* 16,2: 76–116.

Homer-Dixon, Thomas, 1994: "Environmental Scarcities and Violent Conflict. Evidence from Cases", in: *International Security,* 19,1: 5–40.

Homer-Dixon, Thomas, 1996: "Strategies for Studying Complex Ecological-Political Systems", in: *Journal of Environment and Development,* 5,2: 132–148.

Howard, Philip; Homer-Dixon, Thomas, 1995: *Environmental Scarcity and Violent Conflict: The Case of Chiapas, Mexico* (Toronto: Project on Environment, Population, and Security, University College, University of Toronto & Washington, DC: American Association for the Advancement of Science).

Huth, Paul K., 1996: *Standing Your Ground. Territorial Disputes and International Conflict* (Ann Arbor, MI: University of Michigan Press).

Kaplan, Robert, 1994: "The Coming Anarchy", in: *Atlantic Monthly,* 273,2: 44–76.

Kelly, Kimberley; Homer-Dixon, Thomas, 1995: *Environmental Scarcity and Violent Conflict: The Case of Gaza* (Toronto: Project on Environment, Population, and Security, University College, University of Toronto & Washington, DC: American Association for the Advancement of Science).

King, Gary; Keohane, Robert O.; Verba, Sidney, 1994: *Designing Social Inquiry. Scientific Inference in Qualitative Research* (Princeton, NJ: Princeton University Press).

Kolko, Gabriel, 1985: *Anatomy of a War. Vietnam, the United States, and the Modern Historical Experience* (New York: Pantheon).

LaFeber, Walter, 1980: *America, Russia, and the Cold War 1945–1980*, 4th edn. (New York: Wiley). First edition published in 1967.

Lee, Shin-wha, 1996: "Not a One-Time Event: Environmental Change, Ethnic Rivalry, and Violent Conflict in the Third World", Paper presented at the 37th Annual Convention of the International Studies Association, San Diego, CA, 16–20 April.

Levy, Marc A., 1995: "Is the Environment a National Security Issue?", in: *International Security,* 20,2: 35–62.

Libiszewski, Stephan, 1992: "What Is an Environmental Conflict?", in: *Occasional Paper,* no. 1, July (Bern: Swiss Peace Foundation & Zürich: Center for Security Studies and Conflict Research, Swiss Federal Institute of Technology).

Lipschutz, Ronnie D.; Holdren, John P., 1990: "Crossing Borders: Resource Flows, the Global Environment, and International Security", in: *Bulletin of Peace Proposals,* 21,2: 121–133.

Lodgaard, Sverre, 1992: "Environmental Security, World Order, and Environmental Conflict Resolution", in: Gleditsch, Nils Petter (Ed.): *Conversion and the Environment. Proceedings of a Seminar in Perm, Russia, 24–27 November 1991.* PRIO Report, no. 2, May, 115–136.

Lonergan, Steve, 1997: "Water Resources and Conflict: Examples from the Middle East", in: Gleditsch, Nils Petter (Ed.): *Conflict and the Environment* (Dordrecht: Kluwer): 374–384.

Maddox, John, 1972: *The Doomsday Syndrome* (New York: McGraw-Hill).

Mandel, Robert, 1980: "Roots of the Modern Interstate Border Dispute", in: *Journal of Conflict Resolution,* 24,3: 427–434.

McMichael, Anthony J., 1993: *Planetary Overload* (Cambridge: Cambridge University Press). [Quotation is from Canto paperback edition, 1995.]

Meadows, Donella; Meadows, Dennis; Randers, Jørgen; Behrens, William, 1972: *The Limits to Growth: A Report for the Club of Rome's Project on the Predicament of Mankind* (New York: Universe Books).

Midlarsky, Manus I., (Ed.), 1989: *Handbook of War Studies* (Boston, MA: Unwin Hyman).

Midlarsky, Manus I., 1995: "Environmental Influences on Democracy: Aridity, Warfare, and a Reversal of the Causal Arrow", in: *Journal of Conflict Resolution,* 39,2: 224–262.

Mueller, John, 1989: *Retreat from Doomsday. The Obsolescence of Major War* (New York: Basic Books). [Paperback version with a new preface, 1990.]

Muller, Edward N.; Weede, Erich, 1990: "Cross National Variation in Political Violence. A Rational Action Approach", in: *Journal of Conflict Resolution,* 34,4: 624–651.

Myers, Norman; Simon, Julian, 1994: *Scarcity and Abundance? A Debate on the Environment* (New York & London: Norton).

Nordhaus, William D., 1994: *Managing the Global Commons. The Economics of Climate Change* (Cambridge, MA: MIT Press).

North, Robert C., 1990: *War, Peace, Survival. Global Politics and Conceptual Synthesis* (Boulder, CO & London: Westview).

Opschoor, Johannes B., 1989: "North-South Trade, Resource Degradation and Economic Security", in: *Bulletin of Peace Proposals,* 20,2: 135–142.

Percival, Valerie; Homer-Dixon, Thomas, 1995: *Environmental Scarcity and Violent Conflict: The Case of South Africa* (Toronto: Project on Environment, Population, and Security, University College, University of Toronto & Washington, DC: American Association for the Advancement of Science. Revised version (1998) in *Journal of Peace Research,* 35,3: 279–298).

Raknerud, Arvid; Hegre, Håvard, 1997: "The Hazard of War. Reassessing the Evidence for the Democratic Peace", in: *Journal of Peace Research,* 34,4: 385–404.

Renner, Michael, 1996: *Fighting for Survival. Environmental Decline, Social Conflict, and the New Age of Insecurity* (New York & London: Norton, for Worldwatch Institute).

Renner, Michael; Pianta, Mario; Franchi, Cinzia, 1991: "International Conflict and Environmental Degradation", in: Väyrynen, Raimo (Ed.): *New Directions in Conflict Theory. Conflict Resolution and Conflict Transformation* (London: SAGE), in association with the International Social Science Council, 108–128.

Richardson, Lewis Fry, 1960: *Statistics of Deadly Quarrels* (Pittsburgh, PA: Boxwood/Chicago, IL: Quadrangle).

Russett, Bruce, 1993: *Grasping the Democratic Peace. Principles for a Post-Cold War World* (Princeton, NJ: Princeton University Press).

Rønnfeldt, Carsten F., 1997: "Three Generations of Environment and Security Research", in: *Journal of Peace Research*, 34,4: 473–482.

Simon, Julian L., 1989: "Lebensraum: Paradoxically, Population Growth May Eventually End Wars", in: *Journal of Conflict Resolution*, 33,1: 164–180.

Simon, Julian L., 1996: *The Ultimate Resource 2* (Princeton, NJ: Princeton University Press).

Small, Melvin; Singer, J. David, 1982: *Resort to Arms. International and Civil Wars 1816–1980* (Beverly Hills, CA: SAGE).

Smith, Dan; Østreng, Willy (Eds.), 1997: *Research on Environment, Poverty and Conflict: A Proposal, PRIO Report no. 3* (Oslo: International Peace Research Institute, Oslo & the Fridtjof Nansen Institute).

Soroos, Marvin S., 1997a: *The Endangered Atmosphere. Preserving a Global Commons* (Columbia, SC: University of South Carolina Press).

Soroos, Marvin S., 1997b: "The Turbot War: Resolution of an International Fishery Dispute", in: Gleditsch, Nils Petter (Ed.): *Conflict and the Environment* (Dordrecht: Kluwer): 235–252.

Tir, Jaroslav; Diehl, Paul F., 1998: Demographic Pressure and Interstate Conflict: Linking Population Growth and Density to Military Disputes and Wars, 1930–89", in: *Journal of Peace Research*, 35,3: 319–339.

Vasquez, John A., 1993: *The War Puzzle* (Cambridge: Cambridge University Press).

Vasquez, John A., 1995: "Why Do Neighbors Fight? Proximity, Interaction, or Territoriality", in: *Journal of Peace Research*, 32,3: 277–293.

Wallensteen, Peter; Sollenberg, Margareta, 1997: "Armed Conflicts, Conflict Termination and Peace Agreements 1989–96", in: *Journal of Peace Research*, 34,3: 339–358.

Westing, Arthur H. (Ed.), 1986: *Global Resources and International Conflict: Environmental Factors in Strategic Policy and Action* (Oxford: Oxford University Press, for SIPRI).

Westing, Arthur H. (Ed.), 1990: *Environmental Hazards of War. Releasing Dangerous Forces in an Industrialized World* (London: SAGE, for PRIO and UNEP).

Westing, Arthur H. (Ed.), 1997: *Armed Conflict and Environmental Security.* Special issue of *Environment & Security*, 1(2).

Wright, Quincy, 1965: *A Study of War*, 2nd edn. (Chicago, IL: University of Chicago Press). First published in 1942.

Chapter 7
Double-Blind but More Transparent

Journal of Peace Research has now introduced 'double-blind' or 'masked' review procedures. In other words, the author's name and affiliation are removed from the manuscript.
This article explains why we make this change now, why we did not make it before, and
why the decision was not obvious. The main argument in favor of blinding is that the
reviewer should judge the article on the basis of its merit rather than on the basis of the prior
reputation or record of the author. However, the empirical evidence whether or not blinding
makes any difference is mixed, and the practice varies greatly among quality journals. We
make this change mainly because double-blind seems to be the accepted standard among
journals that cater to the same readers and authors, and because we do not want there to be
any doubt as to the journal's commitment to peer review. At the same time, we reiterate our
commitment to transparency, by permitting referees to sign their reports if they want to, by
letting the authors see all the referee reports, by copying the editorial correspondence to the
reviewers, and by strengthening our data replication policies.

From the beginning of 2002, *Journal of Peace Research* has introduced
double-blind review procedures. That is, not only will the identity of the reviewer
normally be unknown to the author, but we will also keep the name of the author
from the referee.[1]

When *JPR* adopted external peer review in late 1983—before that time the
articles were reviewed only by members of the editorial committee—it was thought
impractical and unnecessary to anonymize the articles. We have always protected
the identity of those reviewers who would like to be anonymous; reviewers have the
option of signing their referee report if they wish to be identified, but they are under
no pressure to do so. Hiding the identity of the author from the referee is a slightly
trickier issue. In many cases, it is quite easy for an experienced reviewer to identify
the author, particularly when an earlier version of the article has been presented at a
major conference. With the increasing posting of papers on conference websites and
personal homepages, and the common software feature of filing the name of the
document creator with the document, it has become even easier for a curious
reviewer to establish the identity of the author. For an author to identify the
reviewer is a great deal more difficult, although one can sometimes have a fair

---

[1]I am grateful to the editorial committee of *JPR* for an interesting discussion on these issues at its
meeting on 11–12 January 2002, to Lars Wilhelmsen for assistance, and to Pehr Enckell, John
Langdon, and Sally Morris for comments and information. This article was first published as an
editorial in *Journal of Peace Research* 39(3): 259–262, 2002.

© The Author(s) 2015
N.P. Gleditsch, *Nils Petter Gleditsch: Pioneer in the Analysis of War
and Peace*, SpringerBriefs on Pioneers in Science and Practice 29,
DOI 10.1007/978-3-319-03820-9_7

guess. Journals that now circulate referee reports in electronic form would do well to take note of the less-known features of their computer software.

The main argument in favor of blinding (or masking) the article to a reviewer is straightforward and quite compelling: the reviewer is asked to judge the article on the basis of its merit rather than on the basis of the prior reputation or record of the author. It is not obvious that a high-status author will necessarily get kinder treatment from anonymous reviewers—some junior scholars enjoy the opportunity of trashing the work of a pillar of the profession, while remaining anonymous themselves. But the idea is simply to avoid irrelevant considerations in the editorial process. Moreover, the argument that removing the author's identity from a manuscript is time-consuming is less relevant in the age of word processing.

Nevertheless, we have felt that since a good proportion of the reviewers were likely to guess the identity of the author, it was better to be certain that they knew. When a review is hostile (or friendly) in excess of its substantive argument, and irrelevant considerations seem to be at work, one can adjust for that in the editorial judgment. Making decisions is, after all, the responsibility of the editor. Outside reviews provide advice, but the editor cannot pass the buck. At the level where the decision is made, the identity of the article's author is known.

Like most academic traditions, peer review is a practice that originated in the natural sciences. The *British Medical Journal* used it over 150 years ago (Lock 1986: 3). But double-blind reviewing is by no means a universal practice.[2] Most medical journals do not use it (Davidoff 1998). Neither do the journals of the Royal Society in the UK,[3] but it is 'a common practice in educational research journals' (Abell 1994: 225). Some management journals are reported to practice a severe form of blind review, where referees are requested to disqualify themselves if they know who the authors are, causing one analyst to speculate that 'only those ignorant of the literature would be able to provide reviews for leading researchers' (Armstrong 1997: 70). My own informal survey and personal experience as an author indicate that in political science and international relations, double-blind reviewing is very much the norm—provided the journal is peer reviewed in the first place.

The guidelines for referees in the *Science Editors' Handbook* published by the European Association of Science Editors take an agnostic position on anonymity generally.[4] The publication manual of the American Psychological Association, a book that does not shy away from detailed instructions to authors and editors, is neutral with regard to masked review (APA 1994: 248). The most substantial

---

[2]Foundations dealing with grant applications play a gate-keeping function similar to that of academic journals. Peer review is common, but I am not aware that applications are ever blinded to the referees.

[3]Pers. comm. from John Langdon, Editorial Coordinator, Royal Society, 6 February 2002. Langdon cites concerns about blinding articles that are similar to those I have discussed above.

[4]'If the question of concealing ... either the author's or the referee's name comes up, the editor ... must be very careful' (Enckell 1999: C6: 1). The author is the Managing Editor of *Oikos*, an ecology journal that does not use masked reviewing (Enckell, pers. comm., 5 February 2002).

evidence on editorial practice is found in a survey of 200 journals from all fields conducted in 2000, which found that only 40 % concealed the author's identity (while 90 % concealed the referee's identity from the author).[5] Among the natural sciences, there was a clear majority (2:1) against blinding the article, while in the social sciences, law, and the humanities, there was an even clearer majority in favor (3:1).[6]

I can only speculate about why double-blind reviewing is more common in the social sciences than in the natural sciences. Perhaps the lack of widely accepted theoretical and methodological paradigms in the social sciences leaves them more exposed to partial and irrelevant judgments. Social scientists may also be more alert than natural scientists to issues of fairness and the social functions of evaluation systems.

There is a small experimental and empirical literature on the effect of blind reviewing, but the evidence is mixed (Armstrong 1997; Lock 1986). Some studies find blind reviews to be fairer, others find little difference, and some have even found that blinding harms quality.

At the end of the day, the strongest argument for introducing double-blind procedures in *JPR* is probably that they are so widely accepted in comparable journals. Any journal that does otherwise will easily be seen as a deviant. We have heard very few objections by authors to our practice, but several reviewers have found it unusual and a few have complained. We cannot exclude the possibility that some authors may have avoided submitting to *JPR* because of our excessive openness. We do not want this issue to raise any doubt about the commitment of *JPR* to peer review and impartial quality control. Therefore, we have decided to make articles anonymous before sending them out for review. This change has already been implemented.

We ask all authors to prepare a separate front page with their name and affiliation. This page will be removed before the manuscript is circulated to reviewers. The brief biographical note, which will be required when a manuscript is accepted for publication, should be on a separate, final page. Authors are welcome to keep self-references to published work or conference papers, but should refer to them in the third person rather than by such phrases as 'our work' or 'we have shown earlier'.

We are as strongly committed to transparency as we are to peer review. For that reason, our standard practice has been to circulate to each referee a copy of our letter to the author and all the referee reports. In this way, the referee can see what

---

[5]ALPSP/EASE (2000: 6, 9). This survey of peer review procedures was conducted on the website of the Association of Learned and Professional Society Publishers www.alpsp.org and appears to have proceeded by self-selection. The representativity of the survey is hard to judge. Journals that use peer review are probably overrepresented (none of the journals responding said that they never used it), but there is no obvious reason why this should bias the results on blind reviewing within this group of journals. A wide variety of fields were represented.

[6]Sally Morris, Secretary-General of the ALPSP, supplied the survey data (pers. comm., 7 February 2002). The data are found on our data replication page www.prio.no/cscw/datasets.

use we have made of his or her input to the editorial process, and in what way that input is similar to or different from that of other referees. On this point, we have actually had quite a bit of feedback from our referees, and it has been overwhelmingly positive. We hope that this openness will contribute to even better reviewing in the future. Although double-blind procedures are a little more cumbersome, we will maintain this practice. We will continue to copy the editorial correspondence to the referees, but we will remove the name of the addressee and take care to write the letter in a way that does not hint at the author's identity.

Another way in which we promote transparency is through our replication policy. Since 1998, we have required that authors of articles with systematic empirical information make their data (and associated programming files) available on the web or in a similar fashion. We have also established our own *JPR* data replication page (at www.prio.no/jpr/datasets), where we provide links to the web addresses where the authors have posted their data. Where the authors do not have a suitable website, we post the data on our own website. As of 1 March 2002, this page contains references to 79 datasets.

The profession has a long way to go before the replication norms are practiced smoothly. Anyone who tests the links on our website will discover that some of them lead nowhere; the author has moved, the web address has changed, or (in a very few cases) the author has changed his or her mind or delayed posting the data. In other cases, the data have been posted but only in a general form. The reader is not privileged to know exactly what subset of data was used, and there is no information on coding procedures or calculating routines. Other journals that profess to have a replication policy—and they include most of the journals that are fairly similar to *JPR* in their approach to world politics—have similar problems (Gleditsch/Metelits/Strand 2003).

We are slowly but deliberately strengthening our replication requirement. Authors are asked to supply the data to the editorial office with the final version of the manuscript. We will make the data available directly from *JPR* if the author's website fails to deliver the goods. We hope eventually to find the resources to inspect the data, codebooks, and log files when submitted to us with a view to making sure that replication is actually possible from what is available. We have not yet seriously entertained the idea that replication data might be made available to referees. But we will monitor the international discussion with a view to keeping *JPR* at the forefront of the replication movement. We do this in the firm conviction that King (1995) was right when he portrayed replication as benefiting not only the profession but also the scholar who makes his or her data available. Having other people use your work is a road to academic recognition and should be encouraged by authors as much as by journals. In a study of citations to *JPR* articles in the period 1991–2001, we have found that articles that provide data are more frequently cited, even when controlling for a number of other relevant factors (Gleditsch/Metelits/Strand 2003). Although our replication policy is primarily designed to serve the discipline as a whole, we hope that authors who are given this extra burden of documentation recognize that it is also likely to serve their own interests.

References

Abell, Sandra K., 1994: "Editorial: Ethical Issues for Authors: The Case of the Blind Review", in: *Journal of Research in Science Teaching*, 31,3: 225–226.

ALPSP/EASE Peer Review Survey, 2000: *Worthing: Association of Learned and Professional Society Publishers*, November. www.alpsp.org/peeref.pdf.

APA, 1994: *Publication Manual of the American Psychological Association* (Washington, DC: APA).

Armstrong, J. Scott, 1997: "Peer Review for Journals: Evidence on Quality Control, Fairness, and Innovation", in: *Science and Engineering Ethics*, 3,1: 63–84.

Davidoff, Frank, 1998: "Masking, Blinding, and Peer Review: The Blind Leading the Blinded", in: *Annals of Internal Medicine*, 128,1: 66–68.

Enckell, Pehr H., 1999: *C6: Guidelines on Good Refereeing Practice, in Science Editors' Handbook* (Guildford: European Association of Science Editors), February (C6.1–3).

Gleditsch, Nils Petter; Metelits, Claire; Strand, Håvard, 2003: "Posting Your Data: Will You Be Scooped or Will You Be Famous?", in: *International Studies Perspectives*, 4,1: 89–97.

King, Gary, 1995: "Replication, Replication", *PS: Political Science & Politics*, 28,3: 444–452. See also symposium in the same issue (443–444; 452–499) and Gary King's own replication data page at http://gking.harvard.edu/data.

Lock, Stephen, 1986: *A Difficult Balance: Editorial Peer Review in Medicine* (Philadelphia, PA: ISI Press). First published: London: Nuffield Provincial Hospitals Trust, 1985.

Chapter 8
The Liberal Moment Fifteen Years On

Fifteen years ago, Charles Kegley spoke of a neoidealist moment in international relations. This article examines how the number of armed conflicts has declined in the decade and a half since Kegley's presidential address and shows that the severity of war has been declining over a period of over six decades. The number of countries participating in war has increased, but this is in large measure due to coalition-building in several recent wars. Overall, there is a clear decline of war. It seems plausible to attribute this to an increase in the three factors identified by liberal peace theorists: democracy, trade, and international organization. Four alternative interpretations are examined: the temporary peace, the hegemonic peace, the unsustainable peace, and the capitalist peace. The article concludes that the latter, while running close to the liberal peace interpretation, also presents the greatest challenge to it. Indeed, we seem to be living in a commercial liberal period rather than a world of neoidealism.[1]

8.1 Introduction

Fifteen years ago, exactly to the day, one of my predecessors as President of the ISA, Charles Kegley, alerted us to what he perceived to be a liberal moment in international relations. Or so I thought until I looked up the published version in *International Studies Quarterly* (Kegley 1993).[2] In fact, Kegley used the term

---

[1]This article was originally published in *International Studies Quarterly* 52(4): 691–712, 2008. It is based on my Presidential Address at the 49th Annual Convention of the International Studies Association, San Francisco, CA, 27 March 2008.

[2]I have been privileged to work in an environment full of bright young scholars, many of whom have let me use our joint work or even pilfer their own. Halvard Buhaug, Han Dorussen, Håvard Hegre, Håvard Strand, and Henrik Urdal deserve special mention here. Thanks to the same people plus Kristian Skrede Gleditsch, Ragnhild Nordås, and Gudrun Østby for commenting on an earlier version and, of course, to the members of the Association for electing me. I would also like to record my gratitude to the Research Council of Norway for support for the research reported here, as well as sponsorship of the reception following the Presidential Address. Sage Publications Ltd and PRIO also contributed to the reception. A PowerPoint presentation accompanied the delivery of this address at the convention. The talents of Siri Camilla Aas Rustad are visible in every slide. The presentation can be found, along with references to the data sources, on www.prio.no/cscw/datasets.

© The Author(s) 2015
N.P. Gleditsch, *Nils Petter Gleditsch: Pioneer in the Analysis of War and Peace*, SpringerBriefs on Pioneers in Science and Practice 29,
DOI 10.1007/978-3-319-03820-9_8

'neoidealist' rather than liberal, and there was a question mark in his title.[3] I cannot tell this particular audience that this proves the value of not destroying a good story by checking your sources. But I will stick with my own version for the time being, and return to the question of idealism at the end.

Whether under the heading of idealism or liberalism, it was quite visionary 15 years ago to talk about an emerging international order that might give us a more humane and peaceful world. The Cold War had just ended. But rather than producing peace in Europe, this had reopened old wounds in the Balkans and in the Caucasus. The long-standing armed conflicts in Northern Ireland, Kurdistan, and the Basque territory remained unsolved. In addition, Romania, Moldova, Turkey, Nagorno-Karabakh, Azerbaijan, Chechnya, Moscow, and three places in Georgia saw armed conflict over just a few years at the end of the Cold War.

Three armed conflicts broke out in former Yugoslavia, including the war in Bosnia, which was the bloodiest in Europe since the Greek civil war in the late 1940s. Srebrenica, with the murder of over 7,000 Bosnian men, was still two years down the road (Brunborg et al. 2003). Realists were warning that this was merely the beginning. We were going 'back to the future' or perhaps more appropriately, forward to the past. French-German rivalry would once again play up. The Germans were advised to acquire nuclear weapons to deter the French force de frappe (Mearsheimer 1990). Several contentious issues had arisen between Russia and newly independent Ukraine—the territory of Crimea, the Russian diaspora, the fate of the Soviet Navy, and last but not least, the nuclear arsenal. The realist advice to Ukraine was to hang on to some of the Soviet nuclear weapons in order to deter the Russians (Mearsheimer 1993). On a smaller scale, trouble was brewing between Hungary and several of its neighbors, which host a Hungarian diaspora some 25 % of the population of Hungary itself—in particular with Romania, which under Ceauşescu had actively persecuted them. A minor 'water war' was foreseen Hungary for and Slovakia over the Gabcikovo-Nagymaros dam project on the Danube. Figure 8.1 displays the actual and potential hotspots.

More generally, the number of ongoing state-based armed conflicts[4] had reached a peak in the two years prior to Kegley's address, with 52 armed conflicts active in 38 countries (Gleditsch et al. 2002; Harbom et al. 2008). The number of new conflicts also peaked in the early 1990s.[5] The world did not look like a peaceful place.

---

[3] Although ISI Web of Science has left out the question mark.

[4] Unless otherwise specified, all data on the number of armed conflicts are from the UCDP/PRIO data set. In line with the terminology of the UCDP (Uppsala Conflict Data Program), "state-based" armed conflicts are conflicts over government and territory with at least 25 battle-related deaths in a given year, between two or more organized parties, at least one of which is a government. Thus, the numbers include interstate wars, extrastate (colonial) wars, and civil wars (including internationalized civil wars), but not one-sided conflicts (genocide, politicide) or nonstate conflicts (communal conflicts, intergroup conflicts where government forces are not a party to the fighting). Cf. www.pcr.uu.se/research/UCDP/.

[5] Harbom et al. (2008) and Elbadawi et al. (2008).

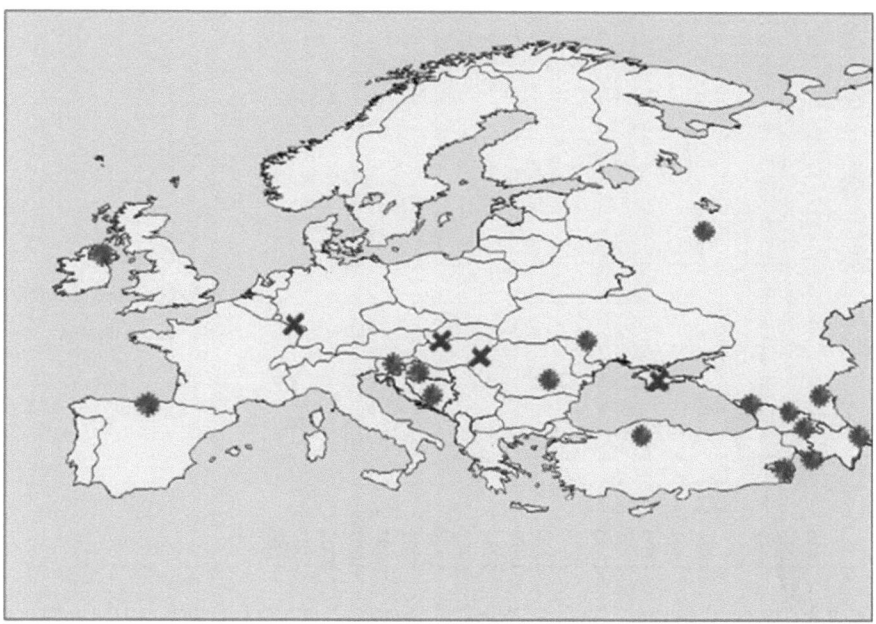

Fig. 8.1 Actual and predicted conflicts in Europe at the end of the Cold War. *Sources* Actual armed conflicts 1989–94 (*) are from the UCDP/PRIO conflict data, cf. Gleditsch et al. (2002) and www.prio.no/cscw/armedconflict. The predicted conflicts (x) are based on my own reading of various media sources at the time. The figure was created by Siri Rustad.

8.2   Less Conflict

No sooner had Charles Kegley announced a possible neoidealist moment than the number of conflicts started to decline, eventually settling to around 30, a level lower than at any time since the mid-1970s (Fig. 8.2).[6] Many conflicts hover just around the threshold of 25 battle deaths per year, so the list of conflicts is not stable from one year to the next, but the number remains about the same. However, after World War II there was a major expansion in the number of independent countries—'the interstate system' in the words of another presidential predecessor, J. David Singer. Internal conflicts in non-independent territories are generally ignored in compilations of armed conflict, so we can easily get an inflated picture of the rise of armed conflict during the Cold War. If we divide the number of armed conflicts by the number of independent countries, we get a much less steep increase up to 1991 and a steeper decline since then, to a level not observed since the early 1960s. The

---

[6]Thirty-four armed conflicts occurred in 25 countries in 2007. The total number of armed conflicts is fairly stable but the 2007 figure stands five above the lowest number (in 2003). But the number of wars (i.e., armed conflicts with more than 1,000 battle deaths in a year), was down to just four in 2007—the lowest in over 50 years.

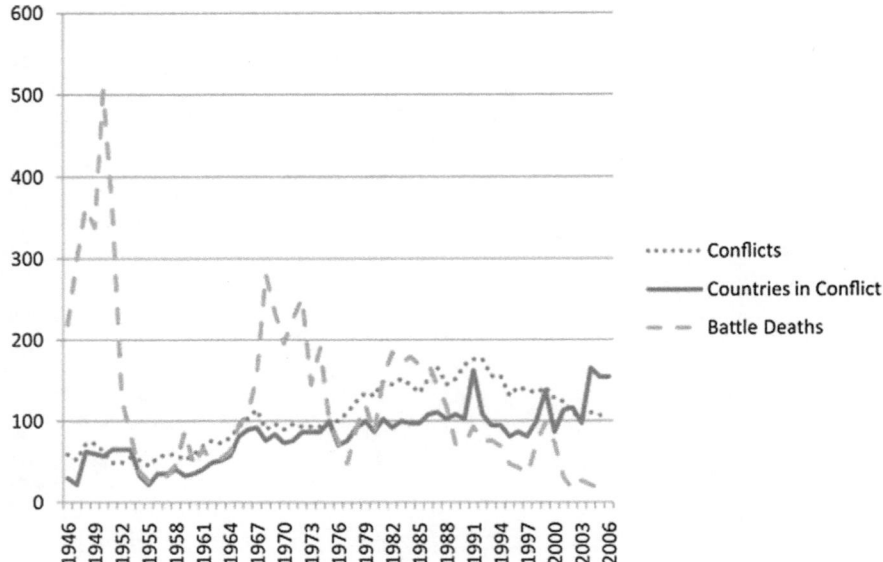

Fig. 8.2 The development of conflict, 1946–2006. *Sources* Number of conflicts and the number of countries participating in conflict based on the UCDP/PRIO data, see Gleditsch et al. (2002) and www.prio.no/cscw/armedconflict. Number of battle deaths (so far updated only until 2005), see Lacina/Gleditsch (2005) and www.prio.no/cscw/cross/battledeaths. The three graphs are set to 100 % in 1975, corresponding to 29 conflicts, 37 countries in conflict, and 135,653 battle deaths. Figure created by Siri Rustad.

number of wars, in the sense of armed conflicts with a minimum of 1,000 battle deaths in a given year, has declined to a quarter of the peak level. There are very few very violent conflicts, but some long-standing low-violence insurgencies persist.[7] In the words of John Mueller (2004), it is mostly 'remnants of war' that are left—opportunistic predation by criminals, bandits, and thugs. The year 2007 was also the fourth year in a row with no recorded interstate conflicts.

Another encouraging sign is that the number of entirely new conflicts has declined even more drastically, to the point where no new conflicts started in 2005 or 2006 (Elbadawi et al. 2008, Fig. 8.2).[8] It is possible, of course, that the remaining conflicts, many of which are decades old, are harder to end than the others. However, as expressed so well in the title of a book by Fred Iklé ([1971] 1991),

---

[7]For other contributions to the argument about 'the waning of war,' see Maoz/Gat (2001) and Väyrynen (2006), in particular the skeptical chapter by Wallensteen (2006). Another skeptical voice is Østerud (2008).

[8]In 2007, two new conflicts started in Niger and DRC, in the sense that they were fought by new actors. However, both countries had experienced armed conflict within the past 10 years.

Every War Must End. This may take time, but as long as a war that ends is not replaced by a new war,[9] we will continue to see a decline in belligerence.

Several alternative measures of conflict show essentially the same picture. For instance, there is also a decline in the share of countries affected by war in their territory. However, one measure appears at first glance to give a different picture. This is the number of countries participating in armed conflict, also plotted in Fig. 8.2. A recent survey of armed conflict (Hewitt et al. 2007) makes a point of this.[10] The main reason why this indicator has not declined and has actually reached a historical peak is the coalition-building in the Gulf War of 1991, the Kosovo War of 1999, and the more recent wars in Afghanistan and Iraq. These four conflicts have from 20 to 36 participants. Compared to the Korean War with 20 participants, and the Vietnam War with just nine participants, these seem like very large wars. At 36 participants, the Iraq War is comparable to the two World Wars using this measure of size. Any other measure, of course, shows otherwise. Many of the participants in the recent wars, such as Iceland or Tonga,[11] have probably joined more as a matter of political solidarity than because they can make a real military contribution. Some of them have suffered no casualties.

Two of these wars, the Gulf War and the Afghanistan War after 2001, were sanctioned by the United Nations. The other two were not, but the United States still went to great lengths to acquire institutional backing from NATO in the case of Kosovo and from a more informal 'coalition of the willing' in the Iraq War. It would certainly be a stretch of imagination to characterize the invasions of Afghanistan in 2001 and Iraq in 2003 as 'peacekeeping.' However, for most of the countries *now* participating in military operations in these countries, the peace-keeping or peacemaking perspective is probably dominant. In that sense the rise in the number of countries nominally 'at war' is a questionable indicator of a resurgence of war. It is more consistent with the concurrent increase in international peacekeeping that we have seen in the same period (Fig. 8.3), an increase almost wholly due to a rise in peacekeeping activities in internal conflicts. The number of personnel participating in such missions has also risen remarkably. But it is hardly reasonable to interpret this as increasing global warfare.

A more pertinent indicator of warfare is the *severity* of war, or the number of people killed in battle. The trend for the 20th century is completely dominated by the two world wars (Lacina et al. 2006: Fig. 8.2). The severity of war clearly peaked in the first half of the 20th century. After World War II the battle deaths continue to be heavily influenced by individual wars, but the peaks are declining (Fig. 8.2). The first peak is the Chinese civil war closely followed by the Korean War; the second is mostly due to the Vietnam War; the third represents the added effects of the

---

[9]Contrary to what is argued in the literature on 'new wars' (Kaldor 1999), these wars are not really new in any important sense (Kalyvas 2001).

[10]Again, if we look at the share of countries participating in armed conflict rather than the number, the curve is less steep.

[11]Tonga joined the multinational force in Iraq on 18 August 2007 (Miles 2007) and thus comes in addition to the 36.

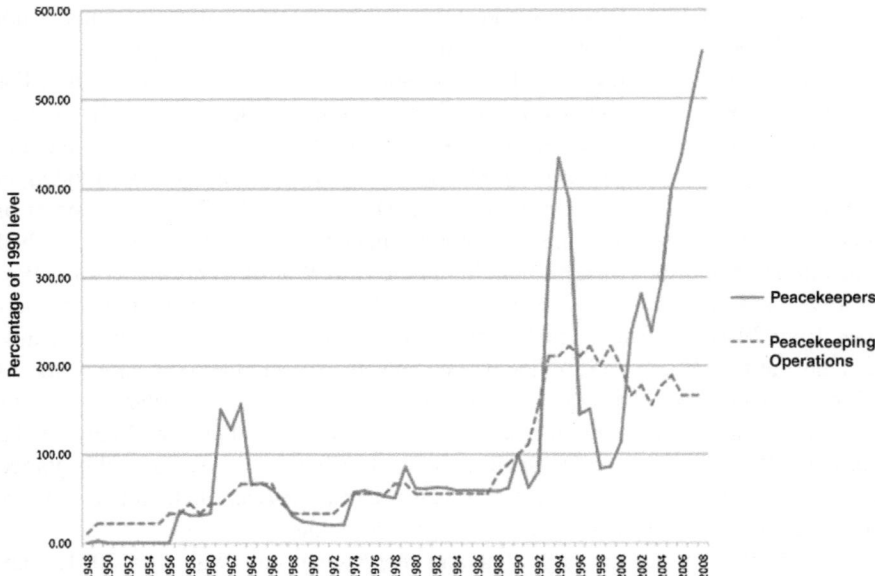

Fig. 8.3 The growth of peacekeeping, 1948–2008. *Sources* Figure created by Siri Rustad based mainly on data from Heldt/Wallensteen (2007), supplemented by Han Dorussen with data from the United Nations. The figures are compiled on a monthly basis, and we have used the January figures for each year

Iran-Iraq War and the Soviet Afghanistan War; and in the fourth, we find the internationalized civil wars of the Democratic Republic of Congo. Figures for war casualties are often highly disputed. The Iraq War is a case in point. An extensive WHO survey (IFHS 2008) puts the number killed at 150,000 for the period March 2003-June 2006.[12] Iraq Body Count (www.iraqbodycount.org), which records published deaths, tallied 48,000 in the same period. The WHO figure probably includes some deaths from types of violence that have not been included in the numbers for earlier wars.[13] But even if we use the WHO figure as battle deaths, they do not reverse the long-term trend toward a lower lethality of war. Currently (Spring 2008), according to the Iraq Body Count's tally of recorded civilian deaths, violence in Iraq seems to be declining. Nevertheless, it remains the most violent enduring war anywhere. Perhaps the Iraq conflict will settle into something like a much more violent version of Northern Ireland. In any case, it is certainly a great tragedy for those involved, as are the conflicts in Darfur, in Colombia, and in the other 30 plus locations. But none of them represents a reversal of the waning of war

---

[12]An earlier survey published in *The Lancet* (Burnham et al. 2006) gives a much higher median estimate of 600,000 for the same period. For a critical comment on the Lancet study and references to the wide-ranging debate about its results, see Munro/Cannon (2008).

[13]The very controversy over the number of civilian deaths in Iraq signals a change in the nature of war reporting and probably in attitudes to war in developed societies.

after the end of the Cold War. In the longer-term perspective, it still seems more appropriate to talk of a world climbing down from a peak of armed violence in the middle of the 20th century.[14]

Statistics of state-based armed conflicts and their consequences in terms of battle-deaths do not tell the full story of human violence. I will leave out violent crime,[15] but deal briefly with three other missing elements. The first concerns *indirect deaths in war.* 'Civil wars kill and maim people—long after the shooting stops,' as another former President of ISA, Bruce Russett, and coauthors said in a seminal article (Ghobarah et al. 2004). Such effects include: revenge killings in the wake of the war; destruction of physical and human capital; slow economic growth; weaker social norms and political chaos; weapons proliferation and crime; increased flows of refugees and internally displaced persons; and environmental destruction including the littering of the landscape with landmines and cluster weapons. A major World Bank study (Collier et al. 2003) aptly characterized civil war as 'development in reverse.' And such consequences do not only affect the countries at war, but frequently also their neighbors.

But such indirect effects of war are not new to our age. The influenza epidemic that followed in the wake of World War I claimed some 40 million lives, more than the war itself (Riley 2001). The war contributed to the spread and lethality of the disease by increasing geographical mobility and by lessening resistance to illness, but we cannot say with any accuracy how many would have died if the war had not contributed. The same problem applies to more recent wars such as those in the DRC and in Sudan. It is possible that indirect effects of war are now greater relative to battle deaths because most armed conflicts take place in poor societies with weak health facilities. But we have no reliable time-series data to back up such a conjecture. The efforts of the international community to ban certain types of weapons, most recently cluster munitions,[16] may also reduce some of these indirect effects in the future.

A second omission is *non-state conflict,* organized groups fighting each other but without the state being a direct party to the conflict. The Uppsala Conflict Data Program records roughly the same number as for state-based conflicts, but they generally involve far fewer fatalities.[17] We do not have comparative data over a long period to establish clear trends, but in sub-Saharan Africa (where most of them occur) both their number and their lethality dropped substantially over the period

---

[14]Terrorism is often portrayed as an exception, but Mack (2008: 1–7) argues persuasively that if Iraq is kept apart (and the killing of civilians in other armed conflicts is usually not included in statistics on terrorism), international terrorism is in fact declining.

[15]Well, not quite. Eisner (2001) shows that crime rates have declined in Europe over several centuries. And Payne (2004) argues that crime as well as war has declined as part of the same civilizing process.

[16]The ban is supported by 111 nations, but its significance is reduced since the opponents include China, India, Israel, Pakistan, Russia, and the United States (Burns 2008).

[17]The UCDP Non-State Conflict Data set covers the period 2002–05 as of 10 August 2008. See Eck/Hultman (2007) and www.pcr.uu.se/research/ucdp/datasets/.

2002–06 (Mack 2008: 36). Given their low severity, they do not appear to offer a serious challenge to the idea of a waning of war.

The third and most serious omission concerns *one-sided conflicts*, that is, genocides, politicides, and, more generally, fatal attacks on unorganized people by governments and other organized groups. Many of these, such as the Holocaust or the liquidation of the Kulaks, are extremely severe and rank with the largest of wars. Most studies focus on individual conflicts or on a short time-range.[18] The best long-range data set is probably the one generated by Rudolph Rummel (1994, 1997) on what he calls democide, defined as 'the murder of any person or people by a government.'[19] It includes 'death by virtue of an intentionally or knowingly reckless ... disregard for life ...' Examples include deadly concentration camps, medical experiments on humans, and famines or epidemics where the authorities withhold aid or even act in a way to make the situation worse. Clearly, Rummel includes deaths that in studies of war would be classified as indirect deaths. The democide data therefore cannot be compared directly with battle deaths in war. But assuming the criteria are reasonably consistent over time, his data show the same inverted U-shaped curve as for battle deaths, peaking in the middle of the 20th century (Rummel 1997: Table 23.1). Given the recent critical examination of China's 'Great Leap Forward' and other disasters under Maoism, it is possible that the peak should be higher, and located later. But the downward trend in recent decades would still hold. The genocides in Rwanda in 1994 and more recently in Darfur, tragic as they are, also do not change the basic shape of the curve.

Tracking all sources of deadly violence is a tall order. Rather than analyzing them one by one or trying to add the number of deaths from the different sources, we may look to life expectancy as a good overall measure of lives not lost. Life expectancy has increased steadily over the last 200 years and is expected to continue to increase for the next half-century (Goklany 2007; Riley 2001; UN 2007). This applies to the world as a whole and to all regions but two: The exceptions are sub-Saharan Africa (mainly because of the AIDS epidemic), and Eastern Europe (because of Russian lifestyle diseases and economic collapse), though the UN projects increasing life expectancies for these regions as well.

The world average for life expectancy has increased from 26 years in the early 19th century to over 65 today (Maddison 2001). In this way, the world has gained many more years of life than lost through war and genocide. And our lives are not only longer, but also healthier (Goklany 2007: 40).

Historians and anthropologists, not to mention archeologists, will be dissatisfied with any reference to data from just the 20th century as 'long-term.' Although we have a large literature on earlier wars, it is difficult to find hard data that would enable us to do systematic comparisons over time. The Correlates of War Project

---

[18]The UCDP One-Sided Violence Data set covers the period 1989–2005 as of 10 August 2008, cf. www.pcr.uu.se/research/ucdp/datasets/. See also Mack (2008: 36).

[19]For more details and discussion, see Rummel's homepage, www.hawaii.edu/powerkills/DBG. CHAP2.HTM. For a critical discussion of some of Rummel's figures, see Dulić (2004a, b) and Rummel (2004).

has taken data on wars and civil wars back to the Congress of Vienna in 1815, but even simple comparisons of the number of wars and casualties become problematic. The first four decades after the Congress of Vienna had very few battle deaths according to the Correlates of War Project, but how meaningful is that information when the interstate system in 1816 consisted of just 23 states (Small/Singer 1982: 47–50)? My immediate predecessor as President of ISA, Jack Levy, has informed us that the number of great-power wars has declined in the last 500 years (Levy et al. 2001: 20), indicating a longer trend in the decline of war.[20] The mass murder of civilians is not a new phenomenon either. Genghis Khan, who is the common ancestor of 0.5 % of all males in the world, according to a recent genetic study (Zerjal et al. 2003), is widely credited with killing over a million people in the Muslim kingdom of Khwarezm in 1220—21.[21]

Judging the long-term development of massacres and wars becomes even more difficult when we move to the pre-historical period. Several anthropologists have argued, in my view convincingly, that the idea of the 'peaceful savage' must be definitively discarded (Gat 2006; Keeley 1996; LeBlanc/Register 2003). Of course, largely peaceful communities can also be found (Fry 2006). War is not intrinsic to human nature, but neither is peace. The decline of violence may be much more of a long-term phenomenon than our statistical data indicate. I am nevertheless inclined to think that the peak of armed violence in the middle of the 20th century is real. We have lived through a particularly lethal combination of the old perception of war as a useful instrument of policy with the modern technological capability to wage war effectively. Our technological skills have continued to improve, so we could kill each other many times over if we applied the full range of human ingenuity to that task. A single direct nuclear exchange between the two superpowers would have changed the picture dramatically and created a more recent and higher peak of war severity. If we do not kill each other at such a rate, it is because our institutions and attitudes have changed. I conclude that we do seem to be moving toward a more peaceful world. But is it a *liberal* peace?

8.3 The Liberal Peace

When Charles Kegley addressed the ISA 15 years ago, the slogan of a liberal peace had not yet been coined, although the key liberal ideas about international relations had reached a venerable age. Karl Deutsch, who surprisingly is not a former President of ISA but whose name adorns one of our awards, had written about international security communities, held together without the use or even the threat of force

---

[20]But in the second half of the 20th century, the bloodiest wars were not direct great-power confrontations, but proxy wars fueled by the two superpowers (Westad 2005).

[21]See Man (2007: 180), who calls 1.25 million 'a conservative estimate'. The widespread story that Genghis Khan killed 1.7 million people in one hour (Clark 2008) should probably be regarded as an urban legend.

(Deutsch et al. 1957). By 1993, a systematic empirical research program on the democratic peace had been under way for over a decade, initiated by Rummel (1983) and Doyle (1983) and later followed up by Maoz/Russett (1993) in particular.[22] The first systematic empirical case for a liberal peace can be dated precisely to 1996 (Oneal et al. 1996) and has been developed at great length in a series of articles and in a book by Russett/Oneal (2001). In fact, the Russett-Oneal project has become one of the most sustained and wide-ranging empirical research efforts to be conducted in any area of international relations. The two have always been very generous in sharing their data, even before IR journals started making this a requirement for publication (Gleditsch 2003). This has led to a number of new challenges to their work. Over the years, then, we have seen improvements in their model, their empirical measures, and their analyses. But they consistently find support for the liberal tripod: democracy, economic integration, and international organization. The literature disagrees on the relative importance of the three factors; some find that given democracy, trade has little importance (Beck et al. 1998), whereas others conclude the opposite (Polachek 1997). But the joint importance of the three factors seems to be well established.[23]

Although the Russett-Oneal work in this area has focused on interstate war, in the tradition of Norman Angell (1910), more recent studies have also established the importance of democracy and trade for civil war (Blanton/Apodaca 2007; Bussmann/Schneider 2007; Hegre et al. 2003).[24] Democracies rarely experience large-scale civil wars[25]; although, some have suffered long and drawn-out secession conflicts at a lower level, particularly where they are fueled by the promise of riches from oil and other raw materials or by religious and ethnic polarization. Much of the popular literature about globalization has emphasized its divisive nature, creating inequalities, and distributional conflict. But most empirical studies show that globalization in fact reduces armed conflict, if not directly then indirectly through its wealth-generating effects. Interestingly, even Katherine Barbieri, one of the first to challenge Oneal and Russett on the interstate liberal peace (Barbieri 1996), has found globalization to reduce the risk of civil war (Barbieri/Reuveny 2005). Since the overwhelming number of conflicts today are internal conflicts, this bodes well for the future of the liberal peace as long as the three liberal factors remain on the rise.

For interstate conflict we cannot as easily generalize from the dyadic to the systemic level. Virtually all the work on the interstate liberal peace is at the dyadic level. Russett and Oneal and others have shown that two countries that share a democratic system trade more, and have more ties through international organizations are less likely to fight. But this does not necessarily mean that a world of more democracies, higher trade, and a proliferation of international organizations

---

[22]The first quantitative study of the democratic peace, hardly noticed at the time, was Babst (1964).

[23]For reviews of this literature, see McMillan (1997), Schneider et al. (2003), and Mansfield/Pollins (2003).

[24]Reports from the State Failure Task Force (Esty et al. 1995, 1998) have shown that economic openness reduces state failure, including internal armed conflict.

[25]Most of the studies of democracy and civil war find an inverted U-shaped relationship (Fearon/Laitin 2003; Hegre et al. 2001). For a recent survey, see Gleditsch et al. (2009).

will produce world peace. In theory, countries could refrain from fighting fellow democracies and their most important trading partners while still maintaining a high level of conflict involvement. In that case the liberal peace would imply a displacement rather than a reduction of warfare. Studies of the systemic effects of democracy provide an ambiguous answer. Of course, if democracies never fight one another, a world with 100 % democracies should have eliminated war. But at lower levels of global democracy the relationship is not so obvious (Gleditsch/Hegre 1997; Mitchell et al. 1999). In a world, or even a region, with no democracy, the emergence of a single democracy might in fact lead to more conflict. The process of democratization itself can also lead to instability and conflict (Mansfield/Snyder 1995; Ward et al. 1998). On the other hand, Mitchell (2002) finds that an increasing proportion of democratic states in the international system promotes the use of democratic norms even by nondemocratic states.

In a rare study of trade relations at the systemic level, Maoz (2006) found that trade interdependence has a consistent dampening effect on the amount of systemic conflict.[26] Lacina et al. (2005) found a statistical relationship between the three liberal factors and the decline in the severity of war, including fatalities in interstate as well as civil wars.[27] However, studies at the system level have few control variables. Generally, the liberal interstate peace is the least well established precisely at the level where it is most important. This has not prevented a range of policy makers from Ronald Reagan to Kofi Annan from embracing the liberal peace, and particularly the democratic peace component. They are probably correct, but it would be reassuring to have more studies at the system level.

Figure 8.4 graphs the development of the three liberal factors over time, normalized around the level in 1973. IGO membership has been increasing almost linearly since the end of World War II. Trade as a share of GDP has also increased most of the time but has exploded since the early 1970s. Finally, democracy has gone through its 'waves' (Huntington 1991) and is now at a level never exceeded, whether measured as the fraction of states under democratic rule or the percentage of world population living in a democracy (Gleditsch/Ward 2006). The rise of the three liberal factors is consistent with the recent decline in the number of wars and the longer decline in the severity of war. But these five curves do not match each other in any simple or convincing way for the entire period since 1945. In the 15 years since Kegley's presidential address, the three liberal factors have gone up and conflict has gone down. But they did not turn around at the same time. The liberal factors were also increasing in the 15 years before his address, while the number of conflicts was rising. The growth of the liberal factors, with a partial exception for democracy, is more consistent with the long-term decline in the lethality of war.

---

[26]Souva/Prins (2006) have also found some evidence for a monadic liberal peace: trade dependence, foreign investment, and democracy reduce a state's propensity to initiate militarized disputes.

[27]Lacina (2006) and Gleditsch et al. (2009) have documented a very clear reduction in the severity of civil war with increasing democracy.

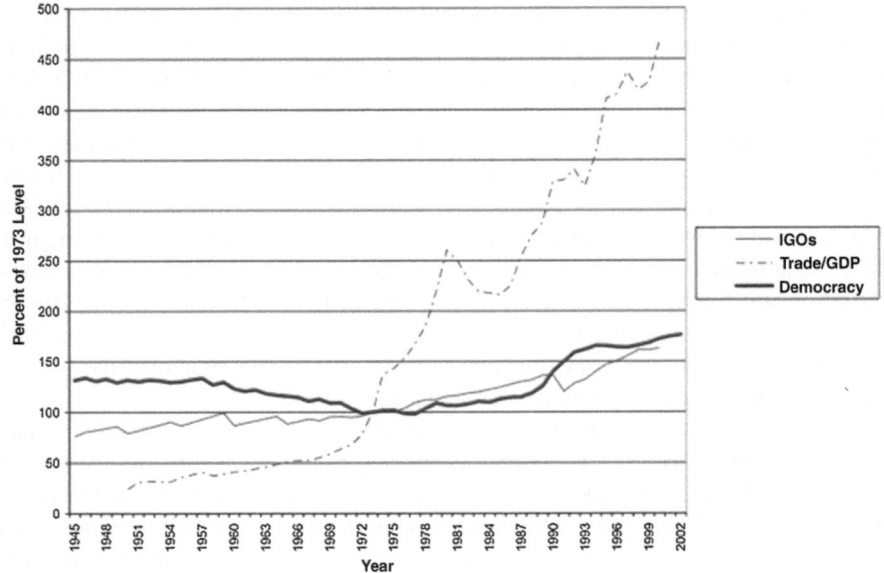

Fig. 8.4 The growth of the liberal factors, 1946–2004. *Source* For each of the four indicators, 1973 = 100 %. *Sources* For data on democracy: Marshall/Jaggers (2003). Polity IV Project, www. systemicpeace.org/, for trade/GDP: Gleditsch (2002), and for IGOs: Pevehouse et al. (2004). Original graph created by Bethany Lacina for Lacina et al. (2005), updated by Kristian Skrede Gleditsch

Despite the widespread acceptance of the idea that the decline of war is related to a liberal peace, there are also a number of alternative interpretations, some of them considerably less optimistic. I will examine four of them here.[28]

8.4 Four Challenges to the Liberal Peace

8.4.1 The Realist Challenge: The Temporary Peace

The major challenge to the liberal peace still comes from the realists. Indeed, Kegley (1993: 143) ended his talk by advocating 'development of a principled realism emphasizing liberal ideals.' For realists, the international system remains anarchic, and its ups and down are determined by the state struggle for survival. Hence the most important variables are the economic and military strength of major powers and

---

[28]Apologies to another predecessor, J. Ann Tickner (1997), for not including a feminist challenge: that the liberal peace is a male peace. I believe, however, that increasing gender inequality rather than challenging the liberal peace will reinforce it. I find some support for that view in the work of Caprioli (2000) and other empirical feminists.

the patterns of alliances. Periods of greater or lesser peace will be explained in terms of the variation of these factors. Realists can argue that the post-World War II period was more peaceful than the previous decades because of its bipolar nature, making a direct confrontation between two blocs armed with nuclear weapons too dangerous. Instead, rivalries were channeled into proxy wars on behalf of the two major blocs. Realism is also sufficiently flexible to account for the decline in global warfare after the Cold War as a result of an even more stable unipolar order, where there is no real challenge to the hegemon. In the words of Fukuyama (1989: 4), democracy and the market economy were 'the final form of government' and thus defined the 'end of history.' But of course the realist factors cannot account in any direct way the ups and downs in the number of conflicts since World War II. Even leaving that aside, the more interesting question is how a real challenge could emerge to the seemingly hegemonic liberal system. It is simple enough to predict the slow relative decline of the United States as the one and only hegemon. Demographic factors and the phenomenal economic growth of China and India dictate that at some point the US economy will be overtaken and other countries will be able to purchase a more powerful military if they so desire. But predicting the slow relative decline of the United States is not the same as predicting the fall of the liberal peace. All the major challengers seem to have embraced the market economy. In three decades, China has moved from being a warfare state, internally as well as externally, to being a trading state, in the words of Richard Rosecrance (1986), another former ISA President. Politically, it has remained a one-party state, with frequent violations of human rights, but without the excesses of the Great Leap Forward or the Cultural Revolution. It seems to be experimenting with competitive elections at the local level and the wisdom of the party leaders is regularly questioned. Corruption is rampant, but corrupt leaders are also regularly being exposed (Thornton 2008). The time when China will be a reliable partner in a democratic peace with its neighbors seems distant, but the economic incentives for maintaining peace appear very robust. Its undemocratic leaders benefit as much from the present trading boom as does the general public, if not more.

Where else can we find a challenge to the hegemony of the market economy? In the remnants of communism in North Korea or Cuba? In the gerontocracy of Zimbabwe? Or among former military coup-makers like Hugo Chavez who can afford to play democrats as long as they can use abundant oil income to boost their position? Such regimes and rulers may have considerable nuisance value to the hegemonic states, but can hardly present a coherent global challenge to the hegemonic system in the same way that communism and fascism did from the 1920s onwards.

The major challenge seems to be found not in traditional economic or military power but in spiritual and cultural power, backed by historical memories. In that sense, it is not a head of state but an opposition leader, Osama bin Laden, who is the main challenger to the international order. Ethno-religious conflict was one of the leading candidates to fill the gap left by the end of the Cold War (Kaplan 1994). Gurr (1994), in the ISA presidential address following Charles Kegley's, referred to a surge in ethnopolitical conflict after the end of the Cold War. However, he did not think there was a strong global force for the further proliferation of such conflicts, and a few years later he proclaimed that ethnic warfare was on the wane (Gurr 2000). Mueller (2000) dismissed the increased concern with ethnic conflict as 'banal.' The general 'clash of civilizations' predicted by Huntington (1993), with the civilizational fault lines largely determined by religion, has hardly been a dominant factor of world politics (Russett et al. 2000) and certainly has not reversed the waning of war. Nevertheless, a number of the major ongoing conflicts, such as the ones in Afghanistan, Iraq, and Sudan, have an important religious element. Indeed, Fig. 8.5 shows that since 1990 an increasing share of the world's armed conflicts have involved Muslim countries, Islamic opposition movements, or both. But this is not due to an absolute increase in what we might call 'Islamic conflicts'; their number remains relatively constant. It is the decline of other types of conflict that creates a relative rise of Islamic conflicts. In other words, in the general trend toward more peace, Muslim countries and Islamic opposition groups seem to be lagging behind, just as Muslim countries in general (and Arab countries in particular) are lagging behind in the rights of women and human rights more generally,[29] in democracy, in the eradication of illiteracy, and in the second demographic transition (UNDP 2005).

This is not a clash of civilizations. Most of these 'Islamic conflicts' pit Islamic opposition movements against the governments of Muslim countries. Although the Iraq War of 2003 was an invasion of a Muslim country by a coalition composed largely of Christian nations, the government of Iraq was a secular, not a religious dictatorship. The Gulf War started because one Muslim country invaded another, as did the Iran-Iraq War. The specter of a mutual crusade or Jihad between Christians and Muslims certainly exists in the minds of many people. We may even fear that it becomes a self-fulfilling prophecy, but it remains a very incomplete description of today's global pattern of conflict. Moreover, there is no evidence that religious conflicts are bloodier than other armed conflicts (Nordås 2007).

Perhaps the greatest realist challenge is what Russett (2005) has called 'bushwacking the democratic peace'. The peace between liberal states has tempted major liberal powers to attempt to help the process along by force. Democracies tend to win the wars in which they participate, and when autocracies lose wars there's a high probability of regime change, which frequently will go in the direction of greater democracy. In a sense, liberal and realist motivations become one and the same. If the West could democratize the Middle East, the liberal audit would be favorable, but so would a traditional security calculation of how to reduce the fear of spreading conflict and the threat to local allies. In this regard, liberals have regarded with some trepidation the lip service paid to the democratic peace by Margaret Thatcher and a series of US Presidents. With the invasion of Iraq in 2003, the worst fears of many liberals seemed to have come to pass. Early empirical work on forced democratization did actually find that military intervention by democracies resulted in some

---

[29]Toft (2007) also finds that Islam was involved in a disproportionate number of civil wars, compared with other religions. However, de Soysa/Nordås (2007) show that Catholic countries scored higher than Muslim countries on political terror in the period 1980–2000.

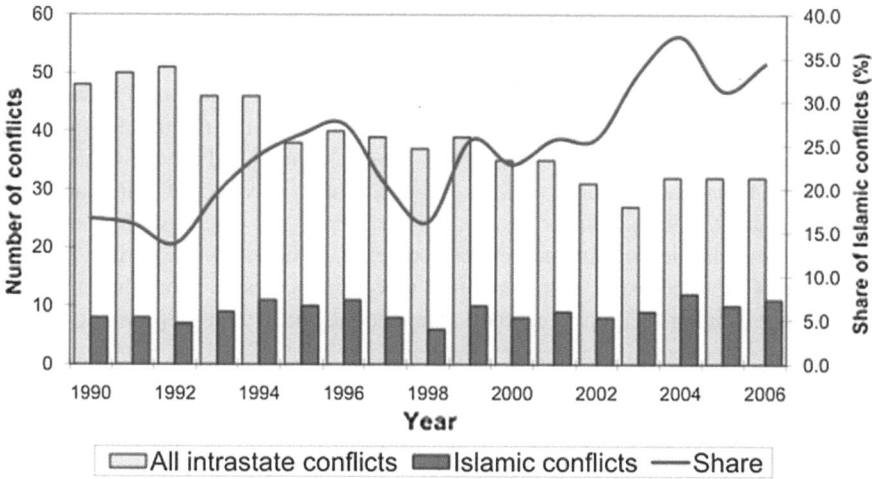

Fig. 8.5 The number and share of armed conflicts involving Muslim countries or Islamic opposition movements or both, 1960–2006. *Source* Figure created by Halvard Buhaug on the basis of the UCDP/PRIO conflict data and his own coding of Islamic conflicts

democratization (Hermann/Kegley 1998; Meernik 1996; Peceny 1999). Research conducted after the Iraq invasion, including that of another former ISA President (Bueno de Mesquita/Downs 2006; see also Pickering/Peceny 2006), has been more skeptical. Forced democratization usually fails to bring the new democracies to a very high level; rather, they tend to end up in the semi-democratic category where political instability and internal conflict is higher than in autocracies (Gleditsch et al. 2007). In the Iraq case, there is an additional reason for skepticism: Even if democratization had been successful, the new democracy would have been surrounded by nondemocracies, a mix for which democratic peace theory does not predict a peaceful future. The only way to overcome that problem would be to extend the policy of forced democratization to Iran, Syria, and others, further strengthening the alliance between liberalism and realism.

One reason why democracies are good at winning wars when they join them is that they are usually more successful than the other side at building alliances, as noted earlier. Even the Iraq War, opposed by many US allies, gathered a 'coalition of the willing' of no less than 36 countries. For liberals, the dilemma is that such coalitions usually include many illiberal states. This is a continuous theme from the Western wartime alliance with Stalin, who killed more people than Hitler, to the US alliance with Saudi Arabia, whose rejection of liberal values is just as firm as those of the enemies of the United States in the region. As Franklin D. Roosevelt reportedly remarked about Rafael Trujillo, dictator of the Dominican Republic, 'He may be an SOB, but he's our SOB.' (Paterson et al. 2005: 157).[30]

---

[30]Similar formulations have been attributed to John Foster Dulles and William Casey. I am grateful to Geir Lundestad for aiding my search.

8.4.2 The Radical Challenge: The Hegemonic Peace

A second challenge to the liberal peace is the radical interpretation. It agrees with the realist view in interpreting the current decline of conflict as mainly a result of an overwhelming hegemony on the part of the West in general and the United States in particular. Thus, the current peace is a hegemonic peace or an imperial peace (Barkawi/Laffey 1999). Unlike realism, the radical view focuses on social and economic inequalities within and between societies. In this view the current peace is the peace of the successfully run plantation where the slaves go about their business without questioning their circumstances. Dependency theory has depicted economic relationships between the center and the periphery as exploitative, where elites on both sides ally against the underdogs (Bornschier/Chase-Dunn 1985; Galtung 1971). Sooner or later, however, the downtrodden are going to rise against domestic elites and international hegemons. In the 1990s, violent street demonstrations in Seattle, Prague, and elsewhere signaled the solidarity of anti-globalization forces with the global underclass. The Marxist states were largely gone, but Marxist analyses of world politics were not.

Much of the anti-globalization literature builds on the premise that in the neo-liberal world, inequalities have been rising within, as well as between, countries. Indeed, Charles Kegley (1993: 140) noted in passing 'the widening gap between the world's rich and poor.' In fact, on a global basis, individual economic inequality has undergone a massive decrease, thanks to the phenomenal economic growth of poor countries like China, India, Vietnam, and many others (Firebaugh 2003; Goklany 2007; Neumayer 2003). During this process inequality within these same countries has increased, not through impoverishment of the masses, but because while wealth is created many people remain left behind in poverty. Thus, we have an unequal peace.[31]

At the global level, Paul Collier's (2007) recent book *The Bottom Billion* argues that the world is making major progress in promoting development, but that one sixth or so of mankind are left out of this process. Around 60 countries suffer not only from low GDP per capita but also low or negative growth. These countries tend to be caught in one or several development traps: armed conflict, the resource curse, being landlocked with bad neighbors, bad governance, or being too small. Unfortunately, Collier will not reveal his list of countries,[32] but by applying his criteria, we find something like the map in Fig. 8.6. By superimposing the ongoing armed conflicts for 2006, we see a certain overlap, but it is by no means perfect.

---

[31]The Centre for Global Research in Canada operates one of many websites www.globalresearch. ca that disseminates such views.

[32]Paul Collier, pers. comm., 26 October 2007. Postscript 2014: The list was released in Collier (2009: 240f).

Fig. 8.6 The Bottom Billion countries and armed conflicts in 2006. *Source* The Bottom Billion data were collected by Åshild Falch. The conflict map was created by Halvard Buhaug and Siri Rustad

Perhaps if the bottom billion notion had taken more account of disparities within countries, there might have been a closer fit.

The literature in political science and economics is divided on the effects of such inequalities. The relative deprivation tradition has pointed to inequality as a cause of conflict (Gurr 1970); although, some have argued that the evidence was inconclusive (Lichbach 1989). Cross-national studies of overall income inequality in a society (so-called vertical inequality) and civil war tend not to find any significant relationship (Collier/Hoeffler 2004; Hegre et al. 2003). But some recent work points to the importance of horizontal inequality in promoting conflict–that is, socioeconomic or political inequalities between groups, such as ethnic or religious groups in generating internal conflict (Østby 2008; Stewart 2002). With increasing inequalities in many countries, this may well be an increasing source of internal conflict. However, it is hardly the stuff of which major wars are made.

8.4.3 The Environmental Challenge: The Unsustainable Peace

Many environmentalists take a dim view of the future and man's exploitation of the natural resource base. This is an old story that goes back to the Malthusian problem of matching food production to population growth. Malthus (1798) thought that this inevitably had to result in a lower birth rate through abortion, infanticide, and birth control (all of which he regarded as sinful) or in a higher death rate through war, famine, and pestilence. In one sense, Malthus was quite correct, since birth control became a widespread way to control population growth, to the point where the UN medium projection for world population shows a leveling-off and even possibly a decline (United Nations 2007). But, of course, attitudes have changed since his time; what Malthus regarded as a sin is now widely regarded as a sensible way to deal with a potential problem. Moreover, food production has increased far beyond the limits that Malthus thought possible.

Despite the seeming irrelevance of the original Malthusian model, neomalthusianism is in many ways the dominant discourse in the public debate on environmental issues. Indeed, Charles Kegley (1993: 140) took for granted that there was an 'unabated deterioration of the global ecosystem.' Neomalthusians argue that we are living on borrowed natural capital, that our ecological footprint is excessive (Wackernagel/Rees 1996), and that at some point the scarcities will become so acute that drastic solutions are inevitable. Paul Ehrlich (1968: 11) announced 40 years ago that 'the battle to feed humanity is over,' and later, Ted Gurr (1985: 51) feared that that overpopulation, exhaustion of nonrenewable energy sources, resource scarcity, and pollution would lead to a crisis of Western societies and more broadly of the whole global system, resulting in greater inequalities, more authoritarianism, and more widespread group conflict.

While Marxism tended toward technological optimism, today's radicalism is to a large extent fused with neomalthusian thought. Thomas Homer-Dixon (1999) describes three forms of environmental scarcity: demand scarcity, supply scarcity, and structural scarcity. The first two relate closely to the original Malthusian model while the third refers to inequality.

While neomalthusian thought is very widespread, it is not unopposed. Cornucopians, or technological optimists, argue that a resource crisis can easily be averted by technological innovation, the substitution of resources, and market pricing. Attitudes change and environmental values start taking precedence over unrestricted economic growth once basic needs have been satisfied.[33] So, as the former Saudi Arabian Minister of Oil Sheik Yamani is reported to have said '… the stone age came to an end not for a lack of stones, and the oil age will end, but not for the lack of oil' (Greider 2000). Of course, if the economic system is flexible enough to adjust gradually to the threat of resource scarcity, there is no need to fight over the scarcities. Therefore, 'water wars' and other major violence resulting from resource scarcity tend to be in the future. There is very little evidence for a general relationship between resource scarcity and civil war (Theisen 2008). Paradoxically, empirical studies show that the higher our ecological footprint, the more peace (Binningsbø et al. 2007). While a 'water war' rhetoric was very common 15 years ago, it has now largely been replaced by an emphasis on the need for cooperation in order to solve the very real problems of lack of clean fresh water in many areas of the world (UNDP 2006). As Riley (2001: 146) argues, Malthus may have been better at summing up the past than predicting the future.

---

[33]For powerful presentations of a more optimistic message, see Goklany (2007), Lomborg (2001), Simon (1996), and—by the Norwegian Prime Minister—Stoltenberg (2006).

Recently, the eco-war perspective has been revived in the debate about climate change. Climate change is indeed a very serious challenge. This is in part because of the accumulating evidence of probable physical effects of global warming, such as changes in precipitation, increasing sea levels, melting of glaciers, and increases in the number of violent weather events. Perhaps even more serious is the uncertainty associated with climate change. The numerous anomalies and deviations from the long-term trends illustrate the problem of making reliable short-term forecasts. They also make it very difficult to design policies for the prevention and mitigation of climate change.

It is evident that climate change will have consequences for human habitation, but using the physical models to derive social effects remains very difficult. It is not surprising, then, that projections for the social and economic effects of climate change are even more controversial than the physical effects. Moreover, while the IPCC summary of the physical effects are based on reviews of thousands of peer-reviewed studies in academic journals, the social effects rest on a much shakier scientific foundation. This is particularly true in assessing the possible security implications of climate change. NGOs, two successive Secretary-Generals of the UN, and numerous national politicians have surmised that climate change is a major security issue, and that we are now seeing the first of many climate wars in Darfur. But although climate change may have exacerbated the relations between herders and farmers in Darfur, area experts cannot disregard the policies of the Sudanese government, the ethnic and religious rivalries, the history of violence in the country, or the role of neighboring conflicts. As for the role of climate change in conflict more generally, there are very few peer-reviewed studies. Indeed, the IPCC (2007) reports are fairly cautious in commenting on the implications for armed conflict. But where they do, they rely on scattered and peripheral sources.

Had I been a neomalthusian addressing the issue of climate change and conflict 15 years ago, I might have been tempted to point out a disturbing covariation over time between global temperature increases and armed conflict, since both had been on the rise. In the most recent decade and a half, the variation is reversed; higher temperatures are associated with less conflict. However, we can deduce very little from two superimposed time trends, although much of the debate about the social effects of man-made climate change is phrased in such terms.

It is not surprising that the apocalyptic nature of the climate change debate should give rise to dystopias like Alan Weisman's (2007) recent book *The World without Us*, where the author concludes that without man earth would be in good shape,[34] and even fringe phenomena like the Voluntary Human Extinction Movement (VHEM) and the Church of Euthanasia, with its four pillars of abortion, suicide, sodomy, and cannibalism. For social scientists, however, a more pressing issue is how earth cannot just survive with man, but even prosper.

---

[34]A movie version, *Life after People*, was aired on US television in January 2008. Other recent films in the same genre include *Cloverfield* and *I am Legend*.

8.4.4 The Commercial Challenge: The Capitalist Peace

The fourth and final challenge to the liberal peace comes from within the liberal school. One of its origins is the old observation that a democracy has never been established in a country that does not have a market economy. In other words, a market economy is a necessary, but by no means a sufficient condition for democracy.

Hegre (2000) has argued that the relationship between trade and conflict is contingent upon development. With increasing economic development, the cost of seizing and holding a territory increases, and the expected utility of conquest decreases. Developed states are therefore more likely to be trading states. Mousseau (2000, 2003a) found that both the democratic peace and the zone of democratic cooperation are substantially limited to economically developed democracies (see also Mousseau et al. 2003). Taking a further step backwards in the causal chain, Weede (1995, 2005) argues that economic freedom, of which free trade is an important component, promotes economic development and thus lays a foundation for democracy and for peace. Mousseau (2003b; see also Mousseau/Mousseau 2008) argues that it is the rise of contractual forms of exchange within a society that accounts for liberal values, democratic legitimacy, and peace among democratic nations. Gartzke (2007) maintains that the existence of market freedom accounts for the effects usually attributed to democracy and trade in analyses of the liberal peace. McDonald (2007) also finds that greater quantities of publicly held assets lead governments to pursue more aggressive foreign policies and increase the likelihood that they will participate in military conflict. Thus, he argues, capitalism promotes peace. Gartzke and Weede both use the term 'the capitalist peace.' There is some disagreement in the literature as to whether the democratic peace should be seen as a mechanism of the capitalist peace or as an alternative theory.

For someone who grew up with the idea that capitalism produces 'merchants of death' (Engelbrecht/Hanighen 1934) who profit from war, the capitalist peace is a difficult notion to swallow. But there would be little point to doing research if all the answers were given ahead of time.[35] I take refuge in the teachings of yet another former President of the ISA, Kenneth Boulding (1989), who distinguished between three forms of power: threat power, economic power, and integrative power. Threat power builds on force and the threat of destruction. Economic power rests on exchange and enlightened self-interest. Integrative power depends on legitimacy, respect, or even love. An actor does something not because he or she is forced to, or even because it is in his or her best interest, but because it is right. The family and many organizations depend mainly on integrative power; although, there may also be elements of force and exchange. Boulding argues that this is the most significant

---

[35]Russett (2009) asserts that Gartzke's attempt to replace democracy with capitalism alone as a cause of peace 'has been refuted', with reference to Dafoe (2008: 1), who writes that 'the notion of a capitalist peace deserves scholarly attention, but it must share the stage with the democratic peace.' The debate continues.

of all forms of power; and in the long run, neither threat power nor exchange power can be upheld without a minimum of legitimacy. Boulding described possible futures rather than predicted one that was most probable. But he certainly viewed a future where integrative power played the major role as the most hopeful scenario for mankind. In the decade and a half since Charles Kegley's Presidential Address, the world has moved in large measure from a threat system to an exchange system. Perhaps in the next 15 years we can discern a clear movement in the direction of an international integrative system. Then, we can really speak of a neoidealist movement in international relations, and I take the liberty of invoking in support of this trend a popular slogan from my own youth: Make love not war! Meanwhile, even if love does not yet govern the world, most of us will probably be pleased that force has been replaced by commerce to such a large extent. Perhaps at this stage we have to make do with a less radical slogan: Make money not war!

References

Angell, Norman, 1910: *The Great Illusion: A Study of the Relation of Military Power in Nations to Their Economic and Social Advantage* (London: Heinemann). Reissued in a new edition, 1934.

Bakst, Dean V., 1964: "Elective Governments—A Force for Peace", in: *Wisconsin Sociologist*, 3,1: 9–14.

Barbieri, Katherine, 1996: "Economic Interdependence: A Path to Peace or a Source of Interstate Conflict?", in: *Journal of Peace Research*, 33,1: 29–49.

Barbieri, Katherine; Reuveny, Rafael, 2005: "Economic Globalization and Civil War", in: *Journal of Politics*, 67,4: 1228–1247.

Barkawi, Tarak; Laffey, Mark, 1999: "The Imperial Peace: Democracy, Force and Globalization", in: *European Journal of International Relations*, 5,4: 403–434.

Beck, Nathaniel; Katz, Jonathan N.; Tucker, Richard, 1998: "Taking Time Seriously: Time-Series-Cross-Section Analysis with a Binary Dependent Variable", in: *American Journal of Political Science*, 42(4): 1260–1288.

Binningsbø, Helga Malmin; de Soysa, Indra; Gleditsch, Nils Petter, 2007: "Green Giant or Straw Man? Environmental Pressure and Civil Conflict, 1961–99", in: *Population and Environment*, 28,6: 337–353.

Blanton, Robert G.; Apodaca, Claire, 2007: "Economic Globalization and Violent Civil Conflict: Is Openness a Pathway to Peace?", in: *Social Science Journal*, 44,4: 599–619.

Bornschier, Volker; Chase-Dunn, Christopher, 1985: *Transnational Corporations and Underdevelopment* (New York: Praeger).

Boulding, Kenneth E., 1989: *Three Faces of Power* (Newbury Park: Sage).

Brunborg, H.; Lyngstad, Torkel H.; Urdal, Henrik, 2003: "Accounting for Genocide: How Many Were Killed in Srebrenica?", in: *European Journal of Population*, 19,3: 229–248.

Bueno de Mesquita, Bruce; Downs, George W., 2006: "Intervention and Democracy", in: *International Organization*, 60,3: 627–649.

Burnham, Gilbert; Lafta, Riyadh; Doocy, Shannon; Roberts, Les, 2006: "Mortality after the 2003 Invasion of Iraq: Cross-Sectional Cluster Sample Survey", in: *Lancet*, 368,9545: 1421–1428.

Burns, John F., 2008: "Britain Joins a Draft Treaty on Cluster Munitions", in: *New York Times*, 29 May.

Bussmann, Margit; Schneider, Gerald, 2007: "When Globalization Discontent Turns Violent: Foreign Economic Liberalization and Internal War", in: *International Studies Quarterly*, 51,1: 79–97.

Caprioli, Mary, 2000: "Gendered Conflict", in: *Journal of Peace Research*, 37,1: 51–68.

Clark, Josh, 2008: Did Genghis Khan Really Kill 1,748,000 People in One Hour? Available at http://people.howstuffworks.com/genghis-khan-murder.htm.

Collier, Paul, 2007: *The Bottom Billion: Why the Poorest Countries Are Failing and What Can Be Done about It* (Oxford: Oxford University Press).

Collier, Paul, 2009: *Wars, Guns & Votes. Democracy in Dangerous Places* (London: Bodley Head).

Collier, Paul; Hoeffler, Anke, 2004: "Greed and Grievance in Civil War", in: *Oxford Economic Papers*, 56,4: 563–595.

Collier, Paul; Elliott, Lani; Hegre, Håvard; Hoeffler, Anke; Reynal-Querol, Marta; Sambanis, Nicholas, 2003: *Breaking the Conflict Trap: Civil War and Development Policy* (New York: Oxford University Press & Washington, DC: World Bank).

Dafoe, Allan, 2008: *Democracy Still Matters: The Risks of Sample Censoring, and Cross-Sectional and Temporal Controls* (Berkeley, CA: Typescript, University of California). For more recent published work by Dafoe on this topic, see www.allandafoe. com.

de Soysa, Indra; Nordås, Ragnhild, 2007: "Islam's Bloody Innards? Religion and Political Terror, 1980–2000", in: *International Studies Quarterly*, 51,4: 927–943.

Deutsch, Karl W.; Burrel, Sidney A.; Kann, Robert A.; Lee, Maurice Jr.; Lichtermann, Martin; Lindgren, Raymond E.; Loewenheim, Francis L.; van Wagenen, Richard W., 1957: *Political Community and the North Atlantic Area* (Princeton, NJ: Princeton University Press).

Doyle, Michael W., 1983: "Kant, Liberal Legacies, and Foreign Affairs", in: *Philosophy & Public Affairs*, Part I, 12,3: 205–235; Part II 12,4: 323–353.

Dulić, Tomislav, 2004a: "Tito's Slaughterhouse: A Critical Analysis of Rummel's Work on Democide", *Journal of Peace Research*, 41(1): 85–102.

Dulić, Tomislav, 2004b: "A Reply to Rummel", in: *Journal of Peace Research*, 41,1: 105–106.

Eck, Kristine; Hultman, Lisa, 2007: "One-Sided Violence Against Civilians in War: Insights from New Fatality Data", in: *Journal of Peace Research*, 44,2: 233–246.

Ehrlich, Paul, 1968: *The Population Bomb* (New York: Ballantine).

Eisner, Manuel, 2001: "Modernization, Self-control and Lethal Violence. The Long-Term Dynamics of European Homicide Rates in Theoretical Perspective", in: *British Journal of Criminology*, 41,4: 618–638.

Elbadawi, Ibrahim; Hegre, Håvard; Milhous, Gary, 2008: "Introduction: The Aftermath of Civil War", in: *Journal of Peace Research*, 45,4: 451–459.

Engelbrecht, Helmuth C.; Hanighen, Frank C., 1934: *Merchants of Death: A Study of the International Armament Industry* (London: Routledge).

Esty, Daniel C.; Goldstone, Jack A.; Gurr, Ted R.; Surko, Pamela T.; Unger, Alan N., 1995: *State Failure Task Force Report* (McLean, VA: Science Applications International Corporation).

Esty, Daniel C.; Goldstone, Jack A.; Gurr, Ted R.; Harff, Barbara; Levy, Marc; Dabelko, Geoffrey; Surko, Pamela T.; Unger, Alan N., 1998: *State Failure Task Force Report: Phase II Findings* (McLean, VA: Science Applications International Corporation).

Fearon, James D.; Laitin, David D., 2003: "Ethnicity, Insurgency, and Civil War", in: *American Political Science Review*, 97,1: 75–90.

Firebaugh, Glenn, 2003: *The New Geography of Global Income Inequality* (Cambridge, MA: Harvard University Press).

Fry, Douglas P., 2006: *The Human Potential for Peace. An Anthropological Challenge to Assumptions about War and Violence* (New York: Oxford University Press).

Fukuyama, Francis, 1989: "The End of History", in: *National Interest*, 16,1: 3–18.

Galtung, Johan, 1971: "A Structural Theory of Imperialism", in: *Journal of Peace Research*, 8,2: 81–117.

Gartzke, Erik, 2007: "The Capitalist Peace", in: *American Journal of Political Science*, 51,1: 166–191.

Gat, Azar, 2006: *War in Human Civilization* (Oxford: Oxford University Press).

Ghobarah, Hazem Adam; Huth, Paul; Russett, Bruce, 2004: "Civil Wars Kill and Maim People— Long after the Shooting Stops", in: *American Political Science Review*, 97,2: 189–202.

Gleditsch, Kristian Skrede, 2002: "Expanded Trade and GDP Data", in: *Journal of Conflict Resolution*, 46,5: 712–724.

Gleditsch, Kristian Skrede; Ward, Michael D., 1999: "A Revised List of Independent States Since 1816", in: *International Interactions*, 25,4: 393–413.

Gleditsch, Kristian Skrede; Ward, Michael D., 2006: "The Diffusion of Democracy and the International Context of Democratization", in: *International Organization*, 60,4: 911–933.

Gleditsch, Nils Petter (Ed.), 2003: "Symposium on Replication in International Studies Research", in: *International Studies Perspectives*, 4,1: 72–107.

Gleditsch, Nils Petter; Christiansen, Lene S.; Hegre, Håvard, 2007: *Democratic Jihad? Military Intervention and Democracy. Working Paper* (15), WSPS 4242 (Washington, DC: World Bank).

Gleditsch, Nils Petter; Hegre, Håvard, 1997: "Peace and Democracy: Three Levels of Analysis", in: *Journal of Conflict Resolution*, 41,2: 283–310.

Gleditsch, Nils Petter; Hegre, Håvard; Strand, Håvard, 2009: "Democracy and Civil War", in: Midlarsky, Manus (Ed.): *Handbook of War Studies III* (Ann Arbor, MI: University of Michigan Press): 155–192 + 301ff.

Gleditsch, Nils Petter; Wallensteen, Peter; Eriksson, Mikael; Sollenberg, Margareta; Strand, Håvard, 2002: "Armed Conflict 1946–2001: A New Dataset", in: *Journal of Peace Research*, 39,5: 615–637.

Goklany, Indur M., 2007: *The Improving State of the World. Why We're Living Longer, Healthier, More Comfortable Lives on a Cleaner Plane* (Washington, DC: CATO Institute).

Greider, William, 2000: "Oil on Political Waters", in: *Nation*, 271,12: 5–6.

Gurr, Ted Robert, 1970: *Why Men Rebel* (Princeton, NJ: Princeton University Press).

Gurr, Ted Robert, 1985: "On the Political Consequences of Scarcity and Economic Decline", in: *International Studies Quarterly*, 29,1: 51–75.

Gurr, Ted Robert, 1994: "Peoples Against States: Ethnopolitical Conflict and the Changing World-System", in: *International Studies Quarterly*, 38,3: 347–377.

Gurr, Ted Robert, 2000: "Ethnic Warfare on the Wane", in: *Foreign Affairs*, 79,3: 52–64.

Harbom, Lotta; Melander, Erik; Wallensteen, Peter, 2008: "The Dyadic Dimensions of Armed Conflict, 1946–2007", in: *Journal of Peace Research*, 45,5: 697–710.

Hegre, Håvard, 2000: "Development and the Liberal Peace: What Does It Take to Be a Trading State?", in: *Journal of Peace Research*, 37,1: 5–30.

Hegre, Håvard; Gissinger, Ranveig; Gleditsch, Nils Petter, 2003: "Globalization and Internal Conflict", in: Schneider, Gerald; Barbieri, Katherine; Gleditsch, Nils Petter (Eds.): *Globalization and Armed Conflict* (Lanham, MD: Rowman & Littlefield): 251–275.

Hegre, Håvard; Ellingsen, Tanja; Gates, Scott; Gleditsch, Nils Petter, 2001: "Toward a Democratic Civil Peace? Democracy, Political Change, and Civil War, 1816–1992", in: *American Political Science Review*, 95(1): 17–33.

Heldt, Birger; Wallensteen, Peter, 2007: *Peacekeeping Operations: Global Patterns of Intervention and Success, 1948–2004* (Stockholm: Folke Bernadotte Academy).

Hermann, Margaret G.; Kegley, Charles W. Jr.; 1998: "The US Use of Military Intervention to Promote Democracy: Evaluating the Record", in: *International Interactions*, 24(2): 91–114.

Hewitt, J. Joseph; Wilkenfeld, Jonathan; Gurr, Ted Robert, 2007: *Peace and Conflict 2008* (Boulder, CO: Paradigm).

Homer-Dixon, Thomas, 1999: *Environment, Scarcity and Violence* (Princeton, NJ: Princeton University Press).

Huntington, Samuel P., 1991: *The Third Wave: Democratization in the Late Twentieth Century* (Norman, OK: Oklahoma University Press).

Huntington, Samuel P., 1993: "The Clash of Civilizations", in: *Foreign Affairs*, 72,3: 22–49.

IFHS, 2008: "Violence-Related Mortality in Iraq from 2002 to 2006", in: *New England Journal of Medicine*, 358,5: 484–493.

Iklé, Fred Charles, 1971: *Every War Must End* (New York: Columbia University Press). [Revised in a new edition, same publisher, 1991.]

IPCC (2007) *Fourth Assessment Report. Climate Change 2007* (Geneva: Intergovernmental Panel on Climate Change), four volumes. www.ipcc.ch/report/ar4/.

Kaldor, Mary, 1999: *New and Old Wars: Organised Violence in a Global Era* (Cambridge: Polity).

Kalyvas, Stathis N., 2001: "'New' and 'Old' Civil Wars—A Valid Distinction?", in: *World Politics*, 54,1: 99–118.

Kaplan, Robert D., 1994: "The Coming Anarchy", in: *Atlantic Monthly*, 2,2: 44–76.

Keeley, Lawrence H., 1996: *War Before Civilization* (New York: Oxford University Press).

Kegley, Charles W., 1993: "The Neoidealist Moment in International-Studies? Realist Myths and the New International Realities", in: *International Studies Quarterly*, 37,2: 131–146.

Lacina, Bethany, 2006: "Explaining the Severity of Civil Wars", in: *Journal of Conflict Resolution*, 50,2: 276–289.

Lacina, Bethany; Gleditsch, Nils Petter, 2005: "Monitoring Trends in Global Combat: A New Dataset of Battle Deaths", in: *European Journal of Population*, 21,2–3: 145–166.

Lacina, Bethany; Russett, Bruce; Gleditsch, Nils Petter, 2005: "The Declining Risk of Death in Battle", Paper Presented at the 46th Annual Convention of the International Studies Association, Honolulu, HI, 2–5 April.

Lacina, Bethany; Gleditsch, Nils Petter; Russett, Bruce, 2006: "The Declining Risk of Death in Battle", in: *International Studies Quarterly*, 50,3: 673–680.

Leblanc, Steven A.; Register, Katherine E., 2003: *Constant Battles: The Myth of the Peaceful, Noble Savage* (New York: St Martin's).

Levy, Jack S.; Walker, Thomas C.; Edwards, Martin S., 2001: "Continuity and Change in the Evolution of Warfare", in: Maoz, Zeev; Gat, Azar (Eds.): *War in a Changing World* (Ann Arbor, MI: University of Michigan Press): 15–48.

Lichbach, Mark I., 1989: "An Evaluation of 'Does Economic Inequality Breed Political Conflict?' Studies", in: *World Politics*, 41,4: 431–470.

Lomborg, Bjørn, 2001: *The Skeptical Environmentalist. Measuring the Real State of the World* (Cambridge: Cambridge University Press).

Mack, Andrew (Ed.), 2008: *Human Security Brief 2007* (Vancouver, BC: Human Security Report Project, Simon Fraser University).

Maddison, Angus, 2001: *The World Economy: A Millennium an Essay on the Principle of Population Perspective* (Paris: OECD).

Malthus, Thomas, 1798: *An Essay on the Principle of Population.* Reprinted (1999) (Oxford: Oxford University Press).

Man, John, 2007: *Genghis Khan: Life, Death, and Resurrection* (New York: St Martin's).

Mansfield, Edward D.; Pollins, Brian 2003: *Economic Interdependence and International Conflict: New Perspectives on an Enduring Debate* (Ann Arbor, MI: University of Michigan Press).

Mansfield, Edward D.; Snyder, Jack, 1995: "Democratization and the Danger of War", in: *International Security*, 20,1: 5–38.

Maoz, Zeev, 2006: "Network Polarization, Network Interdependence, and International Conflict, 1816–2002", in: *Journal of Peace Research*, 43,4: 391–411.

Maoz, Zeev; Russett, Bruce M., 1993: "Normative and Structural Causes of Democratic Peace, 1946–1986", in: *American Political Science Review*, 87,3: 624–638.

Marshall, Monty G., Jaggers, Keith, 2003: *Polity IV Project.* www.systemicpeace.org/.

McDonald, Patrick J., 2007: "The Purse Strings of Peace", in: *American Journal of Political Science*, 51,3: 569–582.

McMillan, Susan M., 1997: "Interdependence and Conflict", in: *Mershon International Studies Review*, 41,1: 33–58.

Mearsheimer, John J., 1990: "Back to the Future—Instability in Europe after the Cold War", in: *International Security*, 15,1: 5–56.

Mearsheimer, John J., 1993: "The Case For a Ukrainian Nuclear Deterrent", in: *Foreign Affairs*, 72,3: 50–66.

Meernik, James, 1996: "United States Military Intervention and the Promotion of Democracy", in: *Journal of Peace Research*, 33,4: 391–401.

Miles, Donna H., 2007: "Pacific Command Chief Praises Little Tonga for Big Iraq Contribution", in: *American Forces Press Service*, 18 September. www.defense.gov/news/newsarticle.aspx? id=47476.

Mitchell, Sara McLaughlin, 2002: "A Kantian System? Democracy and Third-Party Conflict Resolution", in: *American Journal of Political Science*, 46,4: 749–759.

Mitchell, Sara McLaughlin; Gates, Scott; Hegre, Håvard, 1999: "Evolution in Democracy-War Dynamics", in: *Journal of Conflict Resolution*, 43,6: 771–792.

Mousseau, Michael, 2000: "Market Prosperity, Democratic Consolidation, and Democratic Peace", in: *Journal of Conflict Resolution*, 44,4: 472–507.

Mousseau, Michael, 2003a: "An Economic Limitation to the Zone of Democratic Peace and Cooperation", in: *International Interactions*, 28,2: 137–164.

Mousseau, Michael, 2003b: "The Nexus of Market Society, Liberal Preferences, and Democratic Peace: Interdisciplinary Theory and Evidence", in: *International Studies Quarterly*, 47,4: 483–510.

Mousseau, Michael; Hegre, Håvard; Oneal, John R., 2003: "How the Wealth of Nations Conditions the Liberal Peace", in: *European Journal of International Relations*, 9,2: 277–314.

Mousseau, Michael; Mousseau, Demet Yalcin, 2008: "The Socioeconomic Requisite of Liberal Democracy and Human Rights", in: *Journal of Peace Research*, 45,3: 327–344.

Mueller, John, 2000: "The Banality of Ethnic War", in: *International Security*, 25,1: 42–70.

Mueller, John, 2004: *The Remnants of War* (Ithaca, NY: Cornell University Press).

Munro, Neil; Cannon, Carl M., 2008: "Data Bomb", in: *National Journal*, 4 January. www.nationaljournal.com/.

Neumayer, Eric, 2003: "Beyond Income: Convergence in Living Standards, Big Time", in: *Structural Change and Economic Dynamics*, 14,3: 275–296.

Nordås, Ragnhild, 2007: "Are Religious Conflicts Bloodier? Assessing the Impact of Religion on Civilian Conflict Casualties", Paper Presented at the 48th Annual Convention of the International Studies Association, Chicago, IL, 28 February–3 March.

Oneal, John R.; Oneal, Frances H.; Maoz, Zeev; Russett, Bruce, 1996: "The Liberal Peace: Interdependence, Democracy, and International Conflict, 1950–85", in: *Journal of Peace Research*, 33,1: 11–28.

Østby, Gudrun, 2008: "Polarization, Horizontal Inequalities and Violent Civil Conflict", in: *Journal of Peace Research*, 45,2: 143–162.

Østerud, Øyvind, 2008: "Toward a More Peaceful World? A Critical View", in: *Conflict, Security & Development*, 8,2: 233–240.

Paterson, Thomas G.; Cliford, J. Garry; Maddock, Shane J.; Kisatsky, Deborah; Hagan, Kenneth, 2005: *American Foreign Relations*, Vol. 2, 6th edn. (Boston, MA: Houghton Mifflin).

Payne, James L., 2004: *A History of Force. Exploring the Worldwide Movement Against Habits of Coercion, Bloodshed, and Mayhem* (Sandpoint, ID: Lytton).

Peceny, Mark, 1999: "Forcing Them to Be Free", in: *Political Research Quarterly*, 52,3: 549–582.

Pevehouse, Jon; Nordstrom, Timothy; Warnke, K., 2004: *Intergovernmental Organizations, 1815–2000: A New Correlates of War Data Set*. www.correlatesofwar.org.

Pickering, Jeffrey; Peceny, Mark, 2006: "Forcing Democracy at Gunpoint", in: *International Studies Quarterly*, 50,3: 539–559.

Polachek, Solomon W., 1997: "Why Democracies Cooperate More and Fight Less", in: *Review of International Economics*, 5,3: 295–309.

Riley, James C., 2001: *Rising Life Expectancy: A Global History* (Cambridge: Cambridge University Press).

Rosecrance, Richard, 1986: *The Rise of the Trading State. Commerce and Conquest in the Modern World* (New York: Basic Books).

Rummel, Rudolph J., 1983: "Libertarianism and International Violence", in: *Journal of Conflict Resolution*, 27,1: 27–71.

Rummel, Rudolph J., 1994: *Death by Government: Genocide and Mass Murder in the Twentieth Century* (New Brunswick, NJ: Transaction).

Rummel, Rudolph J., 1997: *Statistics of Democide. Genocide and Mass Murder Since 1900* (Charlottesville, VA: Center for National Security Law, School of Law, University of Virginia and Transaction Publishers, Rutgers University).

Rummel, Rudolph J., 2004: "One-Thirteenth of a Data Point Does Not a Generalization Make: A Response to Dulić", in: *Journal of Peace Research*, 41,1: 103–104.

Russett, Bruce, 2005: "Bushwacking the Democratic Peace", in: *International Studies Perspectives*, 6,4: 395–408.

Russett, Bruce, 2009: "Democracy, War, and Expansion Through Historical Lenses", in: *European Journal of International Relations*, 15,1: 9–36.

Russett, Bruce; Oneal, John R., 2001: *Triangulating Peace: Democracy, Interdependence, and International Organizations* (New York: Norton).

Russett, Bruce; Oneal, John R.; Cox, Michaelene, 2000: "Clash of Civilizations, or Realism and Liberalism Déjà Vu? Some Evidence", in: *Journal of Peace Research*, 37,5: 583–608.

Schneider, Gerald; Barbieri, Katherine; Gleditsch, Nils Petter, 2003: "Does Globalization Contribute to Peace? A Critical Survey of the Literature", in: Schneider, Gerald; Barbieri, Katherine; Gleditsch, Nils Petter (Eds.): *Globalization and Armed Conflict* (Lanham, MD: Rowman & Littlefield).

Simon, Julian L., 1996: *The Ultimate Resource 2* (Princeton, NJ: Princeton University Press).

Small, Melvin; Singer, J. David, 1982: *Resort to Arms: International and Civil Wars, 1816–1980* (Beverly Hills, CA: Sage).

Souva, Mark; Prins, Brandon, 2006: "The Liberal Peace Revisited: The Role of Democracy, Dependence, and Development in Militarized Interstate Dispute Initiation, 1950–1999", in: *International Interactions*, 32,2: 183–200.

Stewart, Frances, 2002: Horizontal Inequalities: A Neglected Dimension of Development. *QEH Working Paper* (81) (Queen Elizabeth House: University of Oxford).

Stoltenberg, Jens, 2006, "Tro på framtiden – Et innlegg mot framtidspessimismen [Belief in the Future—The Case against Future Pessimism]", in: *Samtiden*, 3: 90–100.

Theisen, Ole Magnus, 2008: "Blood and Soil? Resource Scarcity and Internal Armed Conflict Revisited", in: *Journal of Peace Research*, 45,6: 801–818.

Thornton, John L., 2008: "Long Time Coming—The Prospects for Democracy in China", in: *Foreign Affairs*, 87,1: 2–22.

Tickner, J. Ann, 1997: "You Just Don't Understand: Troubled Engagements between Feminists and IR Theorists", in: *International Studies Quarterly*, 41,4: 611–632.

Toft, Monica Duffy, 2007: "Getting Religion? The Puzzling Case of Islam and Civil War", in: *International Security*, 31,4: 97–131.

UN, 2007: *World Population Prospects: The 2006 Revision*. ESA/P/ WP.202 (New York: United Nations, Department of Economic and Social Affairs, Population Division). www.un.org/esa/population/publications/wpp2006/wpp2006.htm.

UNDP, 2005: *Arab Human Development Report 2002–2005* (New York: United Nations Development Program and Arab Fund for Social and Economic Development).

UNDP, 2006: *Human Development Report 2006* (Houndsmill: Palgrave Macmillan, for the United Nations Development Programme). http://hdr.undp.org/hdr2006/.

Väyrynen, Raimo (Ed.), 2006: *The Waning of Major War: Theories and Debates* (London & New York: Routledge).

Wackernagel, Mathias; Rees, William E., 1996: *Our Ecological Footprint: Reducing Human Impact on Earth* (Gabriola Island, BC: New Society).

Wallensteen, Peter (2006) "Trends in Major War. Too Early for Waning", in: Väyrynen, Raimo (Ed.): *The Waning of Major War: Theories and Debates* (London: Routledge): 80–93.

Ward, Michael D.; Gleditsch, Kristian Skrede, 1998: "Democratizing for Peace", in: *American Political Science Review*, 92,1: 51–61.

Weede, Erich, 1995: "Economic Policy and International Security: Rent-Seeking, Free Trade and Democratic Peace", in: *European Journal of International Relations*, 1,4: 519–537.

Weede, Erich, 2005: *Balance of Power, Globalization and the Capitalist Peace* (Berlin: Liberal).

Weisman, Alan, 2007: *The World Without Us* (New York: St Martin's).

Westad, Odd Arne, 2005: *The Global Cold War* (Cambridge: Cambridge University Press).

Zerjal, Tatiana et al., 2003: "The Genetic Legacy of the Mongols", in: *American Journal of Human Genetics*, 72(3): 717–721.

Chapter 9
Whither the Weather?

Until recently, most writings on the relationship between climate change and security were highly speculative. The IPCC assessment reports to date offer little if any guidance on this issue and occasionally pay excessive attention to questionable sources. The articles published in this special issue form the largest collection of peer-reviewed writings on the topic to date. The number of such studies remains small compared to those that make up the natural science base of the climate issue, and there is some confusion whether it is the effect of 'climate' or 'weather' that is being tested. The results of the studies vary, and firm conclusions cannot always be drawn. Nevertheless, research in this area has made considerable progress. More attention is being paid to the specific causal mechanisms linking climate change to conflict, such as changes in rainfall and temperature, natural disasters, and economic growth. Systematic climate data are used in most of the articles and climate projections in some. Several studies are going beyond statebased conflict to look at possible implications for other kinds of violence, such as intercommunal conflict. Overall, the research reported here offers only limited support for viewing climate change as an important influence on armed conflict. However, framing the climate issue as a security problem could possibly influence the perceptions of the actors and contribute to a self-fulfilling prophecy.

9.1 Introduction

Violence is on the wane in human affairs, even if slowly and irregularly (Goldstein 2011; Pinker 2011).[1] In recent years, however, pundits and politicians, along with a few scholars, have raised the specter that this encouraging trend towards peace

---

[1]This article was originally published in *Journal of Peace Research* 49(1): 4–9, 2012.

© The Author(s) 2015
N.P. Gleditsch, *Nils Petter Gleditsch: Pioneer in the Analysis of War and Peace*, SpringerBriefs on Pioneers in Science and Practice 29,
DOI 10.1007/978-3-319-03820-9_9

might be reversed by environmental change generally and by climate change specifically.[2] In his acceptance speech for the Nobel Peace prize, for instance, President Obama (2009) warned that '[t]here is little scientific dispute that if we do nothing, we will face more drought, more famine, more mass displacement—all of which will fuel more conflict for decades'. He would have been more accurate had he said that there is little if any scientific agreement about these points.

Despite the increasing certainty about global warming and the man-made contribution to it, the two central premises of the Intergovernmental Panel on Climate Change (IPCC), uncertainty continues about many of the physical consequences of climate change and even more so about the social consequences. This uncertainty is compounded by confusion about the definition of 'climate', an issue to which I return below. The IPCC is not charged with the task of doing research; rather it 'reviews and assesses the most recent scientific, technical and socio-economic information produced worldwide'. In an area where little or no research has been conducted, the IPCC has a poor basis for an assessment. Therefore, the two most recent assessment reports (IPCC 2001, 2007) had little to say about the security implications of climate change. Unfortunately, in the absence of peer-reviewed sources, these reports fell prey to the temptation to cite occasional 'grey material', particularly in the Africa chapter of the 2007 report (Nordås/Gleditsch 2009). Indeed, a document explaining the principles for the preparation of its reports (IPCC 2008) approves the use of non-peer reviewed sources in areas where few peer-reviewed sources are available. In a wide-ranging examination of the IPCC, the InterAcademy Council, an umbrella organization of national academies of science, cited a study that found that while 84 % of the sources for IPCC's Working Group 1 on the physical science basis derived from peer-reviewed sources, it was only 59 % for Working Group 2 on the vulnerability of socio-economic and natural systems to climate change (IAC 2010: 16). It also acknowledged that some governments, particularly in developing countries, had not always nominated the best experts, that the author selection process suffered from a lack of transparency, and that the regional chapters did not always make use of experts from outside the

---

[2]With a single exception (De Stefano et al. 2012) the articles in this special issue are based on papers or presentations at the international conference on 'Climate Change and Security', held in Trondheim, Norway, 21–24 June 2010 under the auspices of the Norwegian Royal Society for Sciences and Letters, on the occasion of its 250th anniversary. A large 'thank you' is due to the Society and its sponsors for the anniversary conferences: NTNU, Statoil/Hydro, and the Norwegian Ministry of Education and Research. Generous additional financial support was provided by the Research Council of Norway. My fellow members of the organizing committee for the conference, Ola Listhaug and Ragnar Torvik, helped to shape the program, raise funds, and get the event off the ground. Rune Slettebak assisted the committee through the whole process, including the selection of conference papers and presentations invited to submit draft articles. Julien Bessière skillfully created and maintained the conference website. We are also grateful to all the participants of the conference and the dozens of reviewers, who have greatly influenced the contents of the special issue. Finally, most of the contributors to the special issue commented critically and constructively on a draft of this introduction, as did Andrew Mack, William Nordhaus, and Roger A. Pielke Jr. None of them share any responsibility for whatever errors remain.

region (IAC 2010: 18)—all of which sheds some light on the discussion of security issues in the Africa chapter in the 2007 report.

In the introduction to the first special issue of an academic journal devoted to the topic of climate change and conflict, Nordås/Gleditsch (2007) found little support for the climate-conflict nexus in the academic literature and outlined five priorities for future research in this area:

- a disentangling of the causal chains between climate change and conflict
- a tighter coupling of climate change models and conflict models
- a reconsideration of the kind of violence expected to result from climate change
- a balance of positive and negative effects
- an increased focus on the Third World where climate change will matter most.

Meanwhile, a number of studies relevant to the climate-conflict nexus have been published and this special issue adds 16 more. What have we achieved in terms of the five goals outlined in 2007?

9.2 Disentangling Causal Chains

Virtually all the articles in this special issue try to disentangle the causal chains between climate change and conflict.[3] By far the largest number of studies in the literature generally and in this issue look at how climate variability and specifically changes in precipitation may affect conflict through adverse effects on rainfed agriculture or cattle herding.[4] Adano et al. (2012: 77), for instance, find for two districts in Kenya that 'more conflicts and killings take place in wet seasons of relative abundance' and Theisen (2012: 93), who also studies Kenya, concludes that 'years following wetter years [are] less safe than drier ones'. Butler/Gates (2012) derive a similar conclusion from a formal model. Benjaminsen et al. (2012: 108) state on the basis of the Mopti region of Mali, at the heart of the Sahel, that there is 'little evidence supporting the notion that water scarcity and environmental change are important drivers of inter-communal conflicts'.[5] Hendrix/Salehyan (2012) conclude on the basis of a new database of social conflict in Africa, that rainfall deviations in either direction are associated with conflict, but that violent events are more responsive to heavy rainfall. Of course, while providing water in abundance, heavy rainfall can also produce subsequent scarcities through the damage caused by flooding. Raleigh/Kniveton (2012), on the basis of data from East Africa, also find that rainfall deviations in either direction are associated with conflict, but argue that

---

[3]For a model of possible causal pathways from climate change to conflict (see Buhaug et al. 2010: Fig. 6).

[4]Although the importance of agriculture is assumed rather than measured in terms of employment or production.

[5]Theisen et al. (2011–12), who use disaggregated data for Africa, also find no relationship between drought and civil war.

civil war is more likely in anomalously dry conditions whereas wet conditions are more likely to be associated with non-state conflict. Koubi et al. (2012) investigate whether climate variability may influence armed conflict through its effect on economic development. Although their literature review leads them to hypothesize that climate variability should affect economic growth, they do not find (either in a global study or in a separate analysis for sub-Saharan Africa) any statistically significant impact of climate variability on growth. There is no general link between climate variability and conflict through economic growth, although autocracies may be more vulnerable to conflict through this mechanism. A few articles also have data on variations in temperature, a possible climate driver of conflict that has received considerable attention in a prominent cross-disciplinary journal (Burke et al. 2009; Buhaug 2010).

Two of the articles here (Slettebak 2012; Bergholt/Lujala 2012) look at natural disasters as a cause of conflict, although the latter article also uses disasters as an instrument for economic shocks. While Slettebak concludes that there may be an increasing trend in climate-related natural disasters, he sharply contradicts earlier research on the link between natural disasters and conflict (e.g. Nel/Righarts 2008) and finds support for an argument derived from crisis sociology that people tend to unite in adversity. Bergholt and Lujala find that natural disasters have a negative effect on economic growth, but that this does not translate into an increased risk of conflict. In a scenario study for sub-Saharan Africa, Devitt/Tol (2012) find that the impact of civil war and climate change on economic growth in Africa has been underestimated.

Despite much public concern about the effects of sea-level rise,[6] this is not yet a theme that has received much attention in the conflict literature. Neither are there any articles on possible adverse security effects of possible countermeasures to climate change—the effect of biofuel on agricultural prices and possibly on food riots could have provided an interesting case.[7]

9.3 Climate Models

Climate research provides an important source of data for much of the research on security effects. The majority of the articles in this issue make use of systematic data on levels and change of precipitation. Most of them use empirical data for the past few decades and assess the empirical regularities that can be assumed to continue at least in the near future. Only two of the articles (Bernauer/Siegfried 2012; De Stefano et al. 2012) cite projections from climate models as well, while Devitt/Tol (2012) use economic projections from IPCC's Special Report on Emission

---

[6]In a wide-ranging review of possible security implications of climate change, Scheffran/Battaglini (2011) include sea-level change as a source of potential conflict in South Asia.

[7]For an ethical argument along these lines, see Gomiero et al. (2010).

Scenarios. While our models of conflict are certainly imperfect, and the ability of social scientists to make predictions is limited (Schneider et al. 2010; Ward et al. 2010), current climate models and even data for the past few decades leave much to be desired in terms of forecasting accuracy and geographical precision.

9.4 Types of Violence

Traditionally, research on armed conflict has concentrated on interstate war and civil war. By far the largest killer in the 20th century, however, was one-sided violence (including genocide and politicide) and environmental change has already been linked by some to major episodes of such violence in Rwanda and Darfur.[8] While so far there is not much evidence that robustly links climate change to major armed conflict of any of these three types, there is a more plausible argument that it may influence intergroup violence below the state level, 'nonstate violence' in the language of the Uppsala Conflict Data Program[9] or intercommunal conflict in the language of Benjaminsen et al. (2012).

The bulk of the articles, however, deal with internal conflict. Although some of them focus exclusively on state-based civil conflicts, others examine non-state conflicts in a rural setting, or both types. None of the articles examine urban conflict or one-sided violence. Five of the articles in this issue examine aspects of interstate conflict, though for the most part at a lower level of violence—militarized disputes rather than major war. Water resources, in the form of shared rivers or aquifers, play a key role in four of these studies. De Stefano et al. (2012) assess the 276 international river basins for changes in water variability and institutional resilience. They map the basins most at risk for hydropolitical tension and discuss how to target capacity-building to strengthen resilience to climate change and the development of mechanisms for cooperation and conflict resolution. Tir/Stinnett (2012) find that water scarcity increases the risk of militarized conflict, but that institutionalized agreements can offset the risk. Bernauer/Siegfried (2012) examine the Syr Darya catchment, a promising candidate for a neo-malthusian conflict over international water resources, but conclude that a militarized interstate dispute is unlikely. Another worst case in terms of the potential for water conflict, the Israeli-Palestinian conflict, is discussed by Feitelson et al. (2012). They conclude that it is unlikely that climate change will directly influence the conflict, although the securitization of the water issue may affect the negotiating positions of the parties.

---

[8]For skeptical discussions of the impact of climate change on the violence in Darfur, see Brown (2010) and Kevane/Gray (2008).

[9]www.pcr.uu.se/research/ucdp/datasets/ucdp_non-state_conflict_dataset_/.

9.5 Balancing Effects?

None of the articles in this issue focus on possible positive effects of climate change. In theory, despite the many pessimistic predictions about global food security under global warming,[10] local or regional improvements in the conditions for food production might offset current food insecurity in some areas and help to lower the risk of local scarcity conflict. But this remains to be studied. Gartzke (2012) argues that economic development, which drives climate change, also lowers the risk of interstate conflict. Therefore, even if climate change drives conflict, the effect may not be visible if it is overshadowed by the peacebuilding effect of economic development. Perhaps the overriding concern with economic development in the Third World can also explain a surprising finding in Kvaløy et al. (2012). Using worldwide public opinion data, they observe widespread concern about global warming, but lower rather than higher in countries that are expected to be more seriously affected.

9.6 Where It Matters?

There is indeed a focus on the developing world. Apart from the articles with a global scope, there is a strong concentration on Africa, particularly south of the Sahara, while one article deals with the Middle East and another with central Asia. The bloodiest wars in the second half of the 20th century occurred in East and Southeast Asia, but by the turn of the century there were fewer conflicts in these areas and those that remained were at much lower levels of severity. The scholarly community may have seen climate-related conflicts as more likely to arise in Africa because of that continent's heavy dependence on rainfed agriculture. But in view of the public concern about the effects of sea-level rise and the melting of the Himalayan glaciers, the impact of climate change for conflict in Asia also seems like a worthwhile topic for future research.

---

[10]IPCC (2007: WG2, Chap. 5, Sect. 5.8.1 Findings and Key Conclusions) concludes with high confidence that '[p]rojected changes in the frequency and severity of extreme climate events will have more serious consequences for food and forestry production, and food insecurity, than will changes in projected means of temperature and precipitation', that '[c]limate change increases the number of people at risk of hunger' but that '[t]he impact of chosen socio-economic pathways (SRES scenario) on the numbers of people at risk of hunger is significantly greater than the impact of climate change', and that '[c]limate change will further shift the focus of food insecurity to sub-Saharan Africa' (so that '[by] 2080, about 75 % of all people at risk of hunger are estimated to live in this region'), and (with medium confidence) that 'moderate warming benefits crop and pasture yields in mid- to high-latitude regions'. Collier et al. (2010) argue that the grave consequences of climate change for agriculture in Africa should be countered by industrialization, urbanization, and new agricultural technology (including genetically modified organisms).

9.7 Other Concerns

Some case study-oriented researchers (Homer-Dixon 1994; Kahl 2006) have argued that many case studies find support for a scarcity model of conflict while large-N statistical research generally fails to do so (see e.g. Theisen 2008). However, other case studies (e.g. Benjaminsen 2008) are closer to the skeptical position. In this issue, and in the current literature generally, there is no systematic difference between case studies and statistical investigations. While some of the case study literature has been criticized for studying only the conflict cases (Gleditsch 1998), it can also be faulted for relatively shallow case description and theoretical myopia. More recently, the large-N conflict literature has moved away from an exclusive reliance on the 'country-year' approach, towards geographical and temporal disaggregation (Cederman/Gleditsch 2009). The ambition is to measure conflict as well as explanatory variables for short time intervals and for subnational regions or territorial grid cells. This approach seems particularly appropriate to the study of effects of variables such as climate change that do not vary along national boundaries, and it may help to bridge the gap between case studies and large-N studies.

One of the lessons that the large-N community could learn from proponents of case studies is the emphasis on interaction effects. Homer-Dixon (1994) and Kahl (2006) do not argue that environmental change generally and climate change specifically have a major impact on conflict—the effect plays out in interaction with exogenous conflict-promoting factors (Buhaug et al. 2008, 2010). Koubi et al. (2012) and Tir/Stinnett (2012) take a step in this direction in testing for interactions with institutions and regime type respectively. Kofi Annan (2006: 9–10) argued in one of his last reports as UN Secretary-General, that 'pollution, population growth and climate change are … occurring now and hitting the poorest and most vulnerable hardest. Environmental degradation has the potential to destabilize already conflict-prone regions, especially when compounded by inequitable access or politicization of access to scarce resources.' Here, he is invoking an interaction effect of climate change with no less than three other variables. Unfortunately, it seems unlikely that case study researchers or large-N scholars will launch a systematic investigation of such complicated interaction patterns any time soon.

In reviewing an article for this issue, William Nordhaus[11] was rather critical: 'this is a paper about weather, not climate'. The Glossary in IPCC (2007) defines climate as 'average weather', usually over a 30-year period.[12] Most of the studies reported here operate over shorter time periods, so this criticism has considerable substance, although Hendrix/Salehyan (2012) and Koubi et al. (2012) measure climate variation as deviations from long-term averages. A few recent studies take a very long-term perspective (e.g. Zhang et al. 2006 for China and Tol/Wagner 2010

---

[11]Review, 16 November 2010; permission to cite by name, personal communication, 4 November 2011.

[12]For a critical discussion of different definitions of climate change, see Pielke (2005).

for Europe). With data for a whole millennium,[13] they conclude that war was more frequent in colder periods. However, Tol and Wagner add that the relationship weakens in the industrialized world. A plausible interpretation of this is that agricultural production suffers in the cold periods, but that with increasing industrialization the world moves away from malthusian constraints. The conflict data used in these studies have not been well tested and for obvious reasons there is a lack of control variables. Based on regularities observed by historians in the distant past and using UCDP/PRIO conflict data for the period 1950–2004, Hsiang et al. (2011) argue that the El Nino/Southern Oscillation (ENSO) has a significant influence on the onset of civil conflict. The link to global warming is tenuous and questions have been raised about the robustness of this finding. But if it holds up, it provides another indication that armed conflict may be related to the climate even in the modern age. In any case, better integration between the long-term climate studies and the studies of 'weather' changes reported here, is a priority item on the research agenda.

9.8 Conclusions

Climate change is the world's first truly global manmade environmental problem[14] and a firm warning that human activities can influence our physical environment on a global scale. The range of possible consequences of climate change is so wide, even for the limited temperature changes foreseen in the IPCC scenarios, that it is difficult to sort out the main priorities. Obviously, if a reversal of the trend towards a more peaceful world was one of these consequences, it should have a prominent place on the policy agenda. Based on the research reported here, such a pessimistic view may not be warranted in the short to medium run. However, as noted by Feitelson et al. (2012) and Salehyan (2008), framing climate change as a security issue may influence the perceptions of the actors in local and regional conflict and lead to militarized responses and thus perhaps contribute to a self-fulfilling prophecy.

The study of the relationship between climate change and conflict has advanced noticeably in the past five years. With regard to how changes in precipitation may influence internal conflict, the one area where we now have a fair number of studies, the dominant view seems to be that rainfall abundance is associated with greater risks than drought and that in any case other conflict-generating factors are more important. Studies of how climate change may promote interstate conflict over water resources also seem to point in the direction of a weak or a null

---

[13]Or even two, as in Zhang et al. (2010).

[14]As distinct from international environmental problems such as transboundary pollution (acid rain, pollution in international rivers). The depletion of the ozone layer was another global problem. But it was solved quite rapidly through a mix of unilateral action and an international agreement, although it will take a few generations for the ozone layer to recover completely.

relationship. In other areas, the number of studies is still very low, so it is premature to offer a summary. On the whole, however, it seems fair to say that so far there is not yet much evidence for climate change as an important driver of conflict. In recent reviews of this literature, Bernauer et al. (2012) and Gleditsch et al. (2011) conclude that although environmental change may under certain circumstances increase the risk of violent conflict, the existing evidence indicates that this is not generally the case.

While we primarily hope that the studies presented here will have an impact on scholarly research in this area, they could also have an influence on policymaking. The IPCC is currently working on its Fifth Assessment Report, scheduled for release in 2013. For the first time, this report will have a chapter on the consequences of climate change for human security, including armed conflict (IPCC, no date). We hope that the studies reported here will contribute to a balanced assessment by the IPCC, built on the best peer-reviewed evidence.

References

Adano, Wario R.; Dietz, Ton; Witsenburg, Karen; Zaal, Fred, 2012: "Climate Change, Violent Conflict and Local Institutions in Kenya's Drylands", in: *Journal of Peace Research*, 49,1: 65–80.

Annan, Kofi, 2006: *Progress Report on the Prevention of Armed Conflict*. Report of the Secretary-General. General Assembly, A/60/891 (New York: United Nations). www.gsdrc.org/go/display&type=Document&id=2813.

Benjaminsen, Tor Arve, 2008: "Does Supply-Induced Scarcity Drive Violent Conflicts in the African Sahel? The Case of the Tuareg Rebellion in Northern Mali", in: *Journal of Peace Research*, 45,6: 819–836.

Benjaminsen, Tor A.; Alinon, Koffi; Buhaug, Halvard; Buseth, Jill Tove, 2012: "Does Climate Change Drive Land-Use Conflicts in the Sahel?", in: *Journal of Peace Research*, 49,1: 97–111.

Bergholt, Drago; Lujala, Päivi, 2012: "Climate-Related Natural Disasters, Economic Growth, and Armed Civil Conflict", in: *Journal of Peace Research*, 49,1: 147–162.

Bernauer, Thomas; Siegfried, Tobias, 2012: "Climate Change and International Water Conflict in Central Asia", in: *Journal of Peace Research*, 49,1: 227–240.

Bernauer, Thomas; Böhmelt, Tobias; Koubi, Vally, 2012: "Environmental Changes and Violent Conflict", in: *Environmental Research Letters*, 7,1: 1–8. doi:10.1088/1748-9326/7/1/015601.

Brown, Ian A., 2010: "Assessing Eco-Scarcity as a Cause of the Outbreak of Conflict in Darfur: A Remote Sensing Approach", in: *International Journal of Remote Sensing*, 31,10: 2513–2520.

Buhaug, Halvard, 2010: "Climate Not to Blame for African Civil Wars", in: *PNAS*, 107,38: 16477–16482.

Buhaug, Halvard; Gleditsch, Nils Petter; Theisen, Ole Magnus, 2008: "Implications of Climate Change for Armed Conflict", in: Paper Prepared for the Social Dimensions of Climate Change Program (Washington, DC: World Bank, Social Development Department). http://siteresources.worldbank.org/INTRANETSOCIALDEVELOPMENT/Resources/SDCCWorkingPaper_Conflict.pdf.

Buhaug, Halvard; Gleditsch, Nils Petter; Theisen, Ole Magnus, 2010: "Implications of Climate Change for Armed Conflict", in: Mearns, Robin; Norton, Andy (Eds.): *Social Dimensions of Climate Change: Equity and Vulnerability in a Warming World* (Washington, DC: World Bank): Chap. 3, 75–101.

Burke, Marshall B.; Miguel, Edward; Satyanath, Shanker; Dykema, John A.; Lobell, David B., 2009: "Warming Increases the Risk of Civil War in Africa", in: *PNAS*, 106,49: 20670–20674.

Butler, Christopher K.; Gates, Scott, 2012: "African Range Wars: Climate, Conflict, and Property Rights", in: *Journal of Peace Research*, 49,1: 23–34.

Cederman, Lars-Erik; Gleditsch, Kristian Skrede, 2009: "Introduction to Special Issue on 'Disaggregating Civil War'", in: *Journal of Conflict Resolution*, 53,4: 487–495.

Collier, Paul; Conway, Gordon; Venables, Tony, 2010: "Climate Change and Africa", in: *Oxford Review of Economic Policy*, 24,2: 337–353.

Devitt, Conor; Tol, Richard S.J., 2012: "Civil War, Climate Change and Development: A Scenario Study for Sub-Saharan Africa", in: *Journal of Peace Research*, 49,1: 129–145.

De Stefano, Lucia; Duncan, James; Dinar, Shlomi; Stahl, Kerstin; Strzepek, Kenneth M.; Wolf, Aaron T., 2012: "Climate Change and the Institutional Resilience of International River Basins", in: *Journal of Peace Research*, 49,1: 193–209.

Feitelson, Eran; Tamimi, Abdelrahman, Rosenthal, Gad, 2012: "Climate Change and Security in the Israeli-Palestinian Context", in: *Journal of Peace Research*, 49,1: 241–257.

Gartzke, Erik, 2012: "Could Climate Change Precipitate Peace?", in: *Journal of Peace Research*, 49,1: 177–191.

Gleditsch, Nils Petter, 1998: "Armed Conflict and the Environment: A Critique of the Literature", in: *Journal of Peace Research*, 35,3: 381–400.

Gleditsch, Nils Petter; Buhaug, Halvard; Theisen, Ole Magnus, 2011: Climate Change and Armed Conflict. Revised Version of a Paper Prepared for the Department of Energy/Environmental Protection Agency Workshop on Research on Climate Change Impacts and Associated Economic Damages, Washington, DC, 27–28 January; published as: Theisen, Ole Magnus; Gleditsch, Nils Petter; Buhaug, Halvard, 2013: "Is Climate Change a Driver of Armed Conflict?", in: *Climate Change*, 117,3: 613–635.

Goldstein, Joshua S., 2011: *Winning the War on War: The Decline of Armed Conflict Worldwide* (New York: Dutton).

Gomiero, Tiziano; Paoletti, Maurizio G.; Pimentel, David, 2010: "Biofuels: Efficiency, Ethics, and Limits to Human Appropriation of Ecosystem Services", in: *Journal of Agricultural & Environmental Ethics*, 23,5: 403–434.

Hendrix, Cullen; Salehyan, Idean, 2012: "Climate Change, Rainfall, and Social Conflict in Africa", in: *Journal of Peace Research*, 49,1: 35–50.

Homer-Dixon, Thomas, 1994: "Environmental Scarcities and Violent Conflict: Evidence from Cases", in: *International Security*, 19,1: 5–40.

Hsiang, Solomon M.; Meng, Kyle C.; Cane, Mark A., 2011: "Civil Conflicts Are Associated with the Global Climate", in: *Nature*, 476(25 August): 438–441.

IAC, 2010: *Climate Change Assessments. Review of the Processes & Procedures of the IPCC* (Amsterdam: InterAcademy Council). http://reviewipcc.interacademycouncil.net/.

IPCC, 2001: *IPCC Third Assessment Report. Climate Change 2001* (Geneva: Intergovernmental Panel on Climate Change). www.ipcc.ch.

IPCC, 2007: *IPCC Fourth Assessment Report. Climate Change 2007* (Geneva: Intergovernmental Panel on Climate Change). www.ipcc.ch; for Appendix I, Glossary, see www.ipcc.ch/publications_and_data/ar4/wg1/en/annexessglossary-a-d.html.

IPCC, 2008: *Appendix A to the Principles Governing IPCC Work* (Geneva: Intergovernmental Panel on Climate Change). First Edition 1999, Last Amended 2008. http://ipcc.ch/pdf/ipcc-principles/ipcc-principles-appendix-a.pdf.

IPCC, no date: Agreed Reference Material for the IPCC Fifth Assessment Report. www.ipcc.ch/pdf/ar5/ar5-outline-compilation.pdf.

Kahl, Colin H., 2006: *States, Scarcity, and Civil Strife in the Developing World* (Princeton, NJ: Princeton University Press).

Kevane, Michael; Gray, Leslie, 2008: "Darfur: Rainfall and Conflict", in: *Environmental Research Letters*, 3,3: 1–10. doi:10.1088/1748-9326/3/3/034006.

Koubi, Vally; Bernauer, Thomas; Kalbhenn, Anna; Spilker, Gabriele, 2012: "Climate Variability, Economic Growth, and Civil Conflict", in: *Journal of Peace Research*, 49,1: 113–127.

Kvaløy, Berit; Finseraas, Henning; Listhaug, Ola, 2012: "The Publics' Concern for Global Warming: A Cross-National Study of 47 Countries", in: *Journal of Peace Research*, 49,1: 11–22.

Nel, Philip; Righarts, Marjolein, 2008: Natural Disasters and the Risk of Violent Civil Conflict", in: *International Studies Quarterly*, 52,1: 159–185.

Nordås, Ragnhild; Gleditsch, Nils Petter, 2007: "Climate Change and Conflict", in: *Political Geography*, 26,6: 627–638.

Nordås, Ragnhild; Gleditsch, Nils Petter, 2013: "The IPCC, Human Security, and the Climate-Conflict Nexus", in: Redclift, Michael; Grasso, Marco (Eds.): *Handbook on Climate Change and Human Security* (London: Elgar): 67–88.

Obama, Barack H., 2009: *Acceptance Speech for the Nobel Peace Prize*. Oslo, 10 December. www.nobelprize.org/nobel_prizes/peace/laureates/2009/obama-lecture.html.

Pielke, Roger A. Jr., 2005: "Misdefining 'Climate Change': Consequences for Science and Action", in: *Environmental Science & Policy*, 8,6: 548–561.

Pinker, Steven, 2011: *The Better Angels of Our Nature* (New York: Viking).

Raleigh, Clionadh; Kniveton, Dominic, 2012: "Come Rain or Shine: An Analysis of Conflict and Climate Variability in East Africa", in: *Journal of Peace Research*, 49,1: 51–64.

Salehyan, Idean, 2008: "From Climate Change to Conflict? No Consensus Yet", in: *Journal of Peace Research*, 45,3: 315–326.

Scheffran, Jürgen; Antonella, Battaglini, 2011: "Climate and Conflicts: The Security Risks of Global Warming", in: *Regional Environmental Change*, 11,1: S27–S39.

Schneider, Gerald; Gleditsch, Nils Petter; Carey, Sabine C., 2010: "Exploring the Past, Anticipating the Future", in: *International Studies Review*, 12,1: 1–7.

Slettebak, Rune T., 2012: "Don't Blame the Weather! Climate-Related Natural Disasters and Civil Conflict", in: *Journal of Peace Research*, 49,1: 163–176.

Theisen, Ole Magnus, 2008: "Blood and Soil? Resource Scarcity and Internal Armed Conflict Revisited", in: *Journal of Peace Research*, 45,6: 801–818.

Theisen, Ole Magnus, 2012: "Climate Clashes? Weather Variability, Land Pressure, and Organized Violence in Kenya, 1989–2004", in: *Journal of Peace Research*, 49,1: 81–96.

Theisen, Ole Magnus; Holtermann, Helge; Buhaug, Halvard, 2011–12: "Climate Wars? Assessing the Claim that Drought Breeds Conflict", in: *International Security*, 36,3: 79–106.

Tir, Jaroslav; Stinnett, Douglas M., 2012: "Weathering Climate Change: Can Institutions Mitigate International Water Conflict?", in: *Journal of Peace Research*, 49,1: 211–225.

Tol, Richard S.J.; Wagner, Sebastian, 2010: "Climate Change and Violent Conflict in Europe over the Last Millennium", in: *Climatic Change*, 99,1–2: 65–79.

Ward, Michael D.; Greenhill, Brian D.; Bakke, Kristin M., 2010: "The Perils of Policy by p-Value: Predicting Civil Conflicts", in: *Journal of Peace Research*, 47,4: 363–375.

Zhang, David D.; Jim, C.Y.; Lin, George C.-S.; He, Yuan-Qing; Wang, James J.; Lee, Harry F., 2006: "Climatic Change, Wars and Dynastic Cycles in China over the Last Millennium", in: *Climatic Change*, 76,3–4: 459–477.

Zhang, Zhibin; Tian, Huidong; Cazelles, Bernard; Kausrud, Kyrre L.; Brauning, Achim; Guo, Fang; Stenseth, Nils Christian, 2010: "Periodic Climate Cooling Enhanced Natural Disasters and Wars in China During AD 10–1900", in: *Proceedings of the Royal Society Part B Biological Sciences*, 277,1701: 3745–3753.

Chapter 10
The Decline of War—The Main Issues

The debate on the waning of war has recently moved into higher gear and this symposium contributes to the debate. This introductory article outlines briefly some of the major issues: nature versus nurture, the reliability of the data, how broadly violence should be defined, whether there is more agreement on the phenomenon than on its causes, and finally whether the future will be like the past.

10.1 The Waning of War Debate

Although several authors have announced a 'waning of war' in recent decades (notably Mueller 1989; Payne 2004), this literature has moved into a higher gear with the recent spate of literature on this topic.[1] Historians (Gat 2006, 2013; Muchembled 2012) and political scientists (Goldstein 2011; Lacina et al. 2006) have entered the fray, along with a science journalist (Horgan 2012), a web designer (Richards 2010)—and, of course, a cognitive psychologist with a massive 800-page tome (Pinker 2011). Despite the breadth of this literature, this is not the end of the argument, but rather the start of a long debate, to which we hope to make a modest contribution with this forum.[2] It includes a statement on human nature and violence (Pinker 2013) and continues with two contributions that are more skeptical to what has become known as the 'declinist' literature (Levy/Thompson 2013; Thayer 2013). Here, I will attempt to summarize some of the main issues that have emerged in the debate so far.

---

[1]This article was part of a symposium (Gleditsch et al. 2013), which originated in a panel at the 54th Annual Convention of the International Studies Association, San Diego, CA, 1–April 2012. I am grateful to Azar Gat, discussant on the panel, for his comments. Useful suggestions were also offered by the journal's editor and referees. My work was supported by the Research Council of Norway. The other contributors to the symposium were Steven Pinker, Bradley A. Thayer, Jack S. Levy, and William Thompson.

[2]Another excellent review is provided in Mack (2013).

© The Author(s) 2015
N.P. Gleditsch, *Nils Petter Gleditsch: Pioneer in the Analysis of War and Peace*, SpringerBriefs on Pioneers in Science and Practice 29,
DOI 10.1007/978-3-319-03820-9_10

10.2 Nature Versus Nurture

Given the extensive polemics against 'the blank slate' in Pinker (2002), one might have expected that he would base his argument relating to violence strongly on changes in human nature. However, Pinker (2011, 2013) argues that human nature continues to hold a potential for violence as well as a potential for peace, and that environmental factors must be taken into account if we are to understand how we curb our tendency to resort to violence. Thayer (2013) agrees with this, although he argues that Pinker underestimates the difficulty in suppressing our inner demons.

10.3 Data

For the recent decline of war, Pinker, Goldstein, and others rely heavily on the UCDP/PRIO data (Gleditsch et al. 2002) and the PRIO battle deaths data (Lacina/Gleditsch 2005). These data are constantly being challenged. For instance, Obermeyer et al. (2008) argued that it is more appropriate to look at a broader set of war deaths, and that there is no evidence of a decline in war deaths over the past 50 years. Spagat et al. (2009), however, found that to reach this conclusion, they had to ignore data for the periods after 1994 and before 1955, base their time trends on extrapolations from a biased convenience sample of only 13 countries, and rely on an estimated constant that is statistically insignificant. Gohdes/Price (2013) argue that while the PRIO battle deaths dataset currently offers the most comprehensive assembly of such data, the information used to establish the dataset is neither sufficient nor of appropriate quality to offer a clear answer as to whether battle deaths have decreased or increased since the end of the Second World War. Lacina/Gleditsch (2013) respond that very strong assumptions must hold in order for measurement errors to explain the trend in battle deaths and conclude that the waning of war is real. This debate is not a sign of weakness. On the contrary, if the data can withstand multiple challenges, our confidence in the real nature of the trend can only increase. Of course, short-term changes like a half-dozen decrease in the number of ongoing armed conflicts in 2010 and a corresponding increase in 2011 (Themnér/Wallensteen 2012) reflect mainly how a number of conflicts hover around the threshold of 25 annual deaths. Such fluctuations should be ignored in the debate about the long-term waning of war.

10.4 What Kinds of Violence Should Be Included?

For a long time, the statistical study of war was focused almost exclusively on interstate war (Singer/Small 1972). While civil war had been the dominant form of conflict in terms of the number of ongoing conflicts since the 1950s, the cross-national study of civil war did not really take off until the late 1990s (Collier/Hoeffler 1998; Hegre et al. 2001; Fearon/Laitin 2003). Despite the high overlap of war and one-sided violence (genocide, politicide), these phenomena have generally been studied separately. Homicide has hardly been studied by conflict scholars. Payne (2004) and Pinker (2011) have broken with this tradition and offer a unified view of human violence, which is even broadened to include painful interrogation, physical punishment of children, and the like. In terms of the overall human propensity for violence, the wider approach seems justified—both one-sided violence and homicide kill more people than wars and civil wars. But, putting too many forms of violence into the same category runs the risk of undermining our ability to find causal explanations, although broad conceptualizations such as 'motive versus opportunity' (Collier/Hoeffler 1998) run through the literatures on international war and civil war as well as crime. Levy/Thompson (2013), however, remain skeptical of our ability to come up with theoretical explanations that will capture the various forms of violence at the different levels of social organization.

10.5 Absolute Numbers or Rates?

A key point in the controversy in the waning of war literature is the use of relative rather than absolute numbers. While World War II certainly claimed more lives than any individual war in the nineteenth century and possibly more lives than in any human-induced disaster ever, its victims made up a smaller fraction of world population than several earlier conflicts. On this basis, the common characterization of the twentieth century as the world's most violent century becomes questionable (Gat 2013). This simple point invites opposition, even anger. The present writer was pointedly asked in a newspaper interview[3] whether a repetition of the Holocaust today would be a smaller crime because world population has more than doubled since that time. Most people would probably agree that the answer to this question is 'no'. On the other hand, the fact that we have not had two or more Holocausts in recent decades can still be seen as an indication of progress toward the reduction in violence. In any case, from World War II until today, the number of people killed in armed conflict and genocide has been in decline, whether you look at absolute or relative figures. More demanding is Thomas Pogge's critique—that human progress is less than its potential.[4] It is certainly arguable that by 2013, we ought to have come further in our reduction in violence as a tool in human affairs. But, it would be hard indeed to establish a baseline over time for our potential to do so.

---

[3]*Klassekampen* (Oslo), 14 July 2012.
[4]Interview in *Klassekampen* (Oslo), 24 July 2012; Pogge (2010).

10.6   More Agreement on Phenomenon than on Explanations

Despite the various critiques, there is wide agreement on the decline of war and other forms of violence. For instance, Levy/Thompson (2013) do not dispute the analyses of trends in war in Pinker (2011), and Thayer (2013) also finds that he 'convincingly demonstrates' the decline in violence. However, the reasons for the decline are less clear. Levy/Thompson (2013: 412) would 'give a greater causal role to material than cultural factors', and Thayer (2013: 405) finds that the cause (or causes) for the decline are less obvious than the decline itself and that 'each major theory of international relation offers an explanation.' In other words, there is not a lack of explanations, but we are unable to choose between alternative plausible theories. That there is greater agreement on the existence of a phenomenon than its explanation is not uncommon in the social sciences, as the literature on the inter-state democratic peace illustrates (see for example, Schneider/Gleditsch 2013). Indeed, even in modern medicine, there is frequently greater agreement on the fact that something works than on why it works.[5]

10.7   Will the Future Be Like the Past?

Even if the trend toward a reduction in violence is accepted, that trend cannot necessarily be extrapolated into the future. Both Thayer (2013) and Levy/Thompson (2013) are skeptical of the declinist thesis for this reason. Thayer argues that there is a lack of 'better angels' outside the West and that even the West may backslide. Moreover, he feels that Pinker underestimates the importance of the international system and the distribution of power. The rise of China is of particular concern. Levy & Thompson argue that a panel in 1912 could have extrapolated from current trends toward a decline of war, completely missing the factors that soon led to World War I.

While Thayer (2013) sees US primacy as a stabilizing force (and the challenge of China as a threat to the stability), almost the direct opposite view is found in 'left' critiques of the declinist view. For instance, Herman/Peterson (2012) argue that world domination by the United States has led to a series of wars and worldwide repression that take the world in the wrong direction and which eventually must lead to a counter-reaction. Another skeptical school of thought is found in environmentalist writing about the destructive effects of environmental change in general and climate change in particular. Pundits and politicians have raised the specter of a warming world ridden by scarcity conflicts, but so far, there is little systematic evidence that points in this direction (Gleditsch 2012; Scheffran et al.

---

[5]See, for instance, http://en.wikipedia.org/wiki/Aspirin. Accessed 29 January 2013.

2012; Theisen et al. 2013). A more optimistic note is struck by Hegre et al. (2013), who find that the factors robustly linked to civil war in the past (such as poverty, ethnic dominance, unfavorable neighborhood) are projected to decrease in the period 2010–50, leading these authors to predict a halving of the proportion of the world's countries that have internal armed conflict.

References

Collier, Paul; Hoeffler, Anke, 1998: "On Economic Causes of Civil War", in: *Oxford Economic Papers*, 50,4: 563–573.

Fearon, James D.; Laitin, David D., 2003: "Ethnicity, Insurgency, and Civil War", in: *American Political Science Review*, 97,1: 75–90.

Gat, Azar, 2006: *War in Human Civilization* (Oxford: Oxford University Press).

Gat, Azar, 2013: "Is War Declining—and Why?", in: *Journal of Peace Research*, 50,2: 149–157.

Gleditsch, Nils Petter, 2012: "Whither the Weather?", in: *Journal of Peace Research*, 49,1: 3–9.

Gleditsch, Nils Petter; Pinker, Steven; Thayer, Bradley A.; Levy, Jack S.; Thompson, William R., 2013: "The Forum: The Decline of War", in: *International Studies Review*, 15,3: 396–419.

Gleditsch, Nils Petter; Wallensteen, Peter; Sollenberg, Margareta; Eriksson, Mikael; Strand, Håvard, 2002: "Armed Conflict 1946–2001", in: *Journal of Peace Research*, 39,5: 615–637.

Gohdes, Anita; Price, Megan, 2013: "First Things First: Assessing Data Quality Before Model Quality", in: *Journal of Conflict Resolution*, 57,6: 1090–1108.

Goldstein, Joshua S., 2011: *Winning the War on War* (New York: Dutton).

Hegre, Håvard; Ellingsen, Tanja; Gates, Scott; Gleditsch, Nils Petter, 2001: "Toward a Democratic Civil Peace? Democracy, Political Change, and Civil War, 1816–1992", in: *American Political Science Review*, 95,1: 33–48.

Hegre, Håvard; Karlsen, Joakim; Nygård, Håvard Mokleiv; Strand, Håvard; Urdal, Henrik, 2013: "Predicting Armed Conflict, 2010–2050", in: *International Studies Quarterly*, 57,2: 250–270.

Herman, Edward S.; Peterson, David, 2012: *Reality Denial* (ColdType e-reader). Available at http://coldtype.net/Assets.12/PDFs/0812.PinkerCrit.pdf. (Last Accessed 29 January 2013.)

Horgan, John, 2012: *The End of War* (San Francisco, CA: McSweeney).

Lacina, Bethany; Gleditsch, Nils Petter, 2005: "Monitoring Trends in Global Combat: A New Dataset on Battle Deaths", in: *European Journal of Population*, 21,2–3: 145–166.

Lacina, Bethany; Gleditsch, Nils Petter, 2013: "The Waning of War is Real: A Response to Gohdes and Price", in: *Journal of Conflict Resolution*, 57,6: 1109–1127.

Lacina, Bethany; Gleditsch, Nils Petter; Russett, Bruce, 2006: "The Declining Risk of Death in Battle", in: *International Studies Quarterly*, 50,3: 673–680.

Levy, Jack S.; Thompson, William R., 2013: "The Decline of War? Multiple Trajectories and Diverging Trends", in: *International Studies Review*, 15,3: 411–416.

Mack, Andrew (Ed.), 2014: *Human Security Report 2013. The Decline in Global Violence: Evidence, Explanation, and Contestation* (Vancouver: Human Security Report Project), www.hsrgroup.org/docs/Publications/HSR2013/HSRP_Report_2013_140226_Web.pdf.

Muchembled, Robert, 2012: *A History of Violence* (Cambridge: Polity [French original 2008]).

Mueller, John, 1989: *Retreat from Doomsday* (New York: Basic Books).

Obermeyer, Ziad; Murray, Christopher J.L.; Gakidou, Emmanuela, 2008: "Fifty Years of Violent War Deaths from Vietnam to Bosnia", in: *British Medical Journal*, 336,7659: 1482–1486.

Payne, James L., 2004: *A History of Force* (Sandpoint, ID: Lytton).

Pinker, Steven, 2002: *The Blank Slate* (New York: Viking).

Pinker, Steven, 2011: *The Better Angels of Our Nature* (New York: Viking).

Pinker, Steven, 2013: "The Decline of War and Conceptions of Human Nature", in: *International Studies Review*, 15,3: 400–405.

Pogge, Thomas, 2010: *Politics as Usual. What Lies Behind the Pro-Poor Rhetoric* (Cambridge: Polity).

Richards, Jesse, 2010: *The Secret Peace* (New York: Book & Ladder).

Scheffran, Jürgen; Brzoska, Michael; Kominek, Jasmin; Link, P. Michael; Schilling, Janpeter, 2012: "Climate Change and Violent Conflict", in: *Science*, 336,6083: 869–871, 18 May.

Schneider, Gerald; Gleditsch, Nils Petter (Eds.), 2013: *Assessing the Capitalist Peace* (London: Routledge).

Singer, J. David; Small, Melvin, 1972: *The Wages of War 1816–1965* (New York: Wiley).

Spagat, Michael; Mack, Andrew; Cooper, Tara; Kreutz, Joakim, 2009: "Estimating War Deaths: An Arena of Contestation", in: *Journal of Conflict Resolution*, 53,6: 934–950.

Thayer, Bradley A., 2013: "Humans, Not Angels: Reasons to Doubt the Decline of War Thesis", in: *International Studies Review*, 15,3: 406–411.

Theisen, Ole Magnus; Gleditsch, Nils Petter; Buhaug, Halvard, 2013: "Is Climate Change a Driver of Armed Conflict?", in: *Climatic Change*, 117,3: 613–625.

Themnér, Lotta; Wallensteen, Peter, 2012: "Armed Conflicts, 1946–2011", in: *Journal of Peace Research*, 49,4: 565–575.

Chapter 11
The IPCC, Conflict, and Human Security

The publication of the report from Working Group II of the Fifth Assessment Report of the Intergovernmental Panel on Climate Change (IPCC) on 31 March 2014 was accompanied by considerable media publicity, some it suggesting that the world was facing an era of violent upheaval. However, the main discussion of conflict, which is found in the chapter on human security, is moderate in tone and cautious in its conclusions. Other chapters in the WG II report use more dramatic language, while a methods chapter completely dismisses the link between climate change and conflict.

The day after the publication of the most recent report from the UN's Intergovernmental Panel on Climate Change (IPCC) on the effects of climate change, the Norwegian daily newspaper *Dagsavisen* was able to report that Norway's Minister for Climate and Environment now envisaged a future world with more conflicts.[1] This is in line with claims made earlier by the Norwegian Nobel Committee. Against this background, I embarked with some anticipation on the report's 2,679 pages. I found that each of the four chapters that address this question gives a slightly different answer.

The question is discussed most thoroughly in the chapter on human security. This IPCC report is the first to contain such a chapter. The report defines human security broadly, in my opinion far too broadly, but a separate sub-section is devoted to violent conflict. This latter subject was mentioned sparsely in the previous two IPCC reports (published in 2001 and 2007), to some extent on the basis of weak sources. This time the scope of the sources is wider, but at the same time more stringent. The chapter concludes that while some studies associate warming and variable precipitation with violent conflict, other studies do not. Accordingly, there is no basis overall for one to conclude that there is a strong connection. This view is consistent with previous summaries of the literature.

The chapter also points out that climate change is generally believed to influence a number of factors that are frequently associated with violent conflict, such as poverty, poor economic growth and misgovernment. This theme recurs in several

---

[1]First published in Norwegian in the Norwegian daily *Aftenposten*, 11 April 2014, www. aftenposten.no/meninger/kronikker/Klimaendringer-og-krig-7535918.html#.U5618yhqPd7. Translation by Fidotext. The English translation was first published at http://blogs.prio.org/2014/04/climate-change-and-war/. A more extensive examination of the IPCC report is found in Gleditsch/Nordås (2014).

© The Author(s) 2015
N.P. Gleditsch, *Nils Petter Gleditsch: Pioneer in the Analysis of War and Peace*, SpringerBriefs on Pioneers in Science and Practice 29, DOI 10.1007/978-3-319-03820-9_11

other chapters. As long as nothing more precise can be said about the strength of the connection in each of the two causal stages (from climate change to risk factors, and from risk factors to violent conflict), we cannot conclude with any certainty about the role of climate change in violent conflict.

In the 2007 report, the chapter on Africa was the source for the most dramatic assertions about possible connections between climate and violent conflict. The tone is more cautious this time. Although there are several references to the possibility of an increase in violent conflict, the report also highlights the disagreement that exists between researchers.

A methodological chapter, under the somewhat dry heading 'Detection and attribution of observed impacts', gives an extremely critical assessment. The chapter builds on all the others, and evaluates whether the available material provides a basis for robust conclusions about a connection between climate change and its presumed effects. The authors point out that all phenomena that are believed to be influenced by climate change are also influenced by many other factors for which it may be difficult to control.

When the authors come to the question of conflict, little is left standing. First, there is a pervasive uncertainty in the literature regarding the empirical findings. And secondly, most empirical research has focused on annual variations in temperature and precipitation instead of deviations from long-term averages. Thus, these studies are more about variations in the weather than about climate change. Hence, this chapter concludes that we cannot say anything certain about the existence or magnitude of climate-change effects on violent conflict. This conclusion is repeated in its entirety three times: first for civil war, then for small-scale communal violence, and finally for violent individual crime.

A chapter on 'emergent risks' is more alarmist. This chapter presents climate change as a potentially significant factor for future conflict. But once again there is an emphasis on the fact that there is 'low confidence' as to the existence of any documented effect of climate change—as oppose to climate variations—on conflict.

I was anxious to see what use the report would make of a controversial article that appeared in *Science* last autumn. That article, which was published just before the IPCC's literature cut-off date, asserted in broad terms that climate in general, and warming in particular, was a significant factor in conflict at all levels, from individual aggression during a heat wave to international warfare and regime collapse on a millennial scale. The article claimed to be the first 'meta-study' of the field, or in other words, the first comparative statistical assessment of results from all relevant previous studies. Most of the authors whose studies were summarized, however, found it hard to recognize the presentation of their own work. A collective response from 26 researchers (including the author of this article) is now in press. But since the response will not appear in print until several months after the IPCC's literature cut-off date, it is obviously not referred to. Nevertheless, the authors of the chapter on human security would have been aware of the debate—among other things, the article was heavily criticized by several leading German climate researchers in a wide-ranging report published in *Der Spiegel* on 1 August last year. In the human security chapter, the controversial *Science* article is treated as one of several summaries of the literature. By contrast, the chapter on 'emergent risks' (which is co-authored by one of the three authors of the *Science* article) presents the article as a contribution with a higher status.

Two summary chapters, the Technical Summary (TS) and the Summary for Policymakers (SPM), contain formulations that are close to those used by the chapter on human security. Regarding conflict, the TS refers to several of the other chapters, but not to the methodological chapter. There is also a significant linguistic nuance in that the TS claims that climate change will increase risks of violent conflict, while the SPM claims that climate change can increase such risks. The use of word such as 'can' or 'may' in academic writing is extremely problematic, as it provides no basis for evaluating the probability of an event, beyond the fact that it is not zero.

The TS and the SPM are no doubt the most politically influential parts of the report. Shortly before the report was made public, it emerged that one of the two coordinating lead authors of the chapter on economic effects, Richard Tol, had withdrawn from further work on the SPM because he thought that the summary articulated a pessimism for which there was no basis in the individual chapters.

The IPCC's view as to the risks of climate change leading to violent conflict thus depends to some extent on the chapter one chooses to rely on. In my opinion, the methods chapter is the most solidly based, but like the authors of last year's controversial *Science* article, I am not an impartial observer. In any event we can be confident in saying that the IPCC report does not put forward a consensus that climate change will lead to more wars.

References

Gleditsch, Nils Petter; Nordås, Ragnhild, 2014: "Conflicting Messages? The IPCC on Conflict and Human Security", in: *Political Geography*, 43(November): 82–90.

Peace Research Institute Oslo

Independent—International—Interdisciplinary

The Peace Research Institute Oslo (PRIO) conducts research on the conditions for peaceful relations between states, groups and people. Researchers at PRIO work to identify new trends in global conflict, as well as to formulate and document new understanding of and response to armed conflict. They seek to understand how people are impacted by, and cope with, armed conflict, and we study the normative foundations of peace and violence.

PRIO's purpose is to conduct research for a more peaceful world. In pursuit of this, the institute cultivates academic excellence, communicate with communities of scholars, policy-makers, practitioners, as well as the general public, and engages in shaping the global peace research agenda.

PRIO strives to be at the cutting edge analytically as well as in the impact of peace research on policy and practice.

About PRIO

Founded in 1959, the Peace Research Institute Oslo (PRIO) is an independent research institution known for its effective synergy of basic and policy-relevant research. PRIO also conducts graduate training and is engaged in the promotion of peace through conflict resolution, dialogue and reconciliation, public information, and policymaking.

People at PRIO

PRIO has an international staff of approximately 75 (counted in person-years), of which more than 50 are researchers, including doctoral candidates. The institute maintains an administrative/support staff of 15. Within the Norwegian setting, PRIO staff stand out for their high levels of professionalism and their academic

© The Author(s) 2015
N.P. Gleditsch, *Nils Petter Gleditsch: Pioneer in the Analysis of War and Peace*, SpringerBriefs on Pioneers in Science and Practice 29, DOI 10.1007/978-3-319-03820-9

productivity. The Institute's governing board consists of five external appointees and two staff members. PRIO is an equal opportunities employer and values staff diversity.

Research at PRIO

Research at the Institute is multidisciplinary and concentrates both on the driving forces as well as the consequences of violent conflict, and on ways in which peace can be built, maintained and spread. Projects carried out at the Institute are organized within thematic research groups, and researchers at PRIO are in addition organized in three administrative departments and the PRIO Cyprus Centre. From 2002 through 2012, PRIO hosted the Centre for the Study of Civil War (CSCW), a long-term, interdisciplinary initiative that was awarded Centre of Excellence status and core funding by the Research Council of Norway. The diversity of disciplines at PRIO creates a thriving research community that attracts both scholars and funding from around the world.

Journals at PRIO

The Institute owns and hosts the editorial offices of two international peer-reviewed journals—*Journal of Peace Research* and *Security Dialogue*—both of which are edited at PRIO and published by Sage Publications in London. In addition, PRIO houses the editors of *International Area Studies Review* and the *Journal of Military Ethics*. The Institute also publishes reports and policy briefs. Institute researchers maintain high levels of productivity in the form of peer-reviewed articles in top international journals and books with reputable academic publishers.

Research and Engagement

At PRIO, academic research and engagement in peace processes go hand in hand: all peacebuilding engagements are rooted in solid research competence and feed into ongoing research—and ultimately to published academic work. The Institute's policy-relevant findings are in high demand among international bodies (the UN, the World Bank), NGOs, the media and governments, including a number of Norwegian ministries.

Oslo and Nicosia

The Institute is located in modern research facilities in central Oslo. It maintains a separate office in Nicosia: the PRIO Cyprus Centre (PCC). The PCC is committed to research and dialogue aimed at contributing to an informed public debate on key issues relevant to an eventual settlement of the Cyprus problem. Researchers attached to the PCC include both Greek Cypriots and Turkish Cypriots.

Economy and Funders

Budgeted turnover for PRIO as a whole in 2015 is approximately 120 million Norwegian kroner (equivalent to roughly €13 million or $16 million). The Institute has a bottom-up and project-based budget model, where all research engagements depend on the acquisition of external funding. PRIO staff are skilled at combining research innovation with project-development initiative. Major sources of funding include the Research Council of Norway, Norwegian government ministries, the European Comission and a variety of international organizations and foundations. Website: www.prio.org/

Norwegian University of Science and Technology, Trondheim

The Norwegian University of Science and Technology (NTNU) is a public research university located in the city of Trondheim, Norway. NTNU is the second largest of the eight universities in Norway, and has the main national responsibility for higher education in engineering and technology. In addition to engineering and the natural and physical sciences, the university offers advanced degrees in other academic disciplines ranging from the social sciences, the arts, medicine, architecture, and fine art.

NTNU was formed in 1996 by the merger of the Norwegian Institute of Technology (NTH), the Norwegian College of General Sciences (AVH), the Museum of Natural History and Archaeology (VM), the Faculty of Medicine (DMF), the Trondheim Academy of Fine Art and the Trondheim Conservatory of Music (MiT). Prior to the 1996 merger, NTH, AVH, DMF, and VM together constituted the University of Trondheim (UNiT), which was a much looser organization. However, the university's roots go back to 1760, with the foundation of the Trondheim Society, which in 1767 became the Royal Norwegian Society of Sciences and Letters. In 2010 the society, and NTNU, as the society's museum now is part of the university, celebrated its 250th anniversary to commemorate this history. NTNU itself celebrated the 100th anniversary of the foundation of NTH in the same year.

NTNU is governed by a board of 11 members. Two of the members are elected by and among the students.

The university consists of seven faculties with a total of 48 departments and has approximately 22,000 students:

- Faculty of Architecture and Fine Art
- Faculty of Engineering Science and Technology
- Faculty of Humanities
- Faculty of Natural Sciences and Technology
- Faculty of Information Technology, Mathematics and Electrical Engineering
- Faculty of Medicine
- Faculty of Social Sciences and Technology Management

Academic and administrative staff contribute 5,100 person-years of which 3,100 are in education and research. NTNU has more than 100 laboratories and is at any

© The Author(s) 2015

165

N.P. Gleditsch, *Nils Petter Gleditsch: Pioneer in the Analysis of War and Peace*, SpringerBriefs on Pioneers in Science and Practice 29, DOI 10.1007/978-3-319-03820-9

time running some 2,000 research projects. Students and staff can take advantage of roughly 300 research agreements or exchange programs with 58 institutions worldwide.

NTNU's overall budget in 2011/2012 was 673 million euros, most of which came from the Norwegian Ministry of Education. Funding from the Research Council of Norway (NFR) totaled 82 million euros.

The university is home to four of 21 Norwegian Centers of Excellence. These are the Centre for Ships and Ocean Structures, the Centre for the Biology of Memory and the Centre for Quantifiable Quality of Service in Communication Systems. The Centre for the Biology of Memory is also one of four Kavli Neuroscience Institutes. In 2012 Prime Minister Jens Stoltenberg opened the Norwegian Brain Centre one of the largest research laboratories of its kind anywhere, an outgrowth of NTNU's Kavli Institute for Systems Neuroscience.

To increase open access publishing, NTNU has established a publishing fund. In 2008 NTNU's digital institutional repository was founded. The intention was to establish a full-text archive for the documentation of the scientific output of the institution, and to make as much as possible of the material available online, both nationally and internationally. In addition to research articles and books, intended for academics and researchers both inside and outside the university, NTNU disseminates news to the public about the institution and its research and results.

NTNU specializes in technology and the natural sciences, but also offers a range of bachelor's, master's and doctoral programmes in the humanities, social sciences, economics and public and business administration, and aesthetic disciplines. The university also offers professional degree programmes in medicine, psychology, architecture, the fine arts, music, and teacher education, in addition to technology.

NTNU had 84,797 applicants in 2011 and a total student population of 19,054, of whom 9,062 were women. There were 6,193 students enrolled in the Faculty of Social Sciences and Technology Management, 3,518 in the Faculty of Engineering Science and Technology, 3,256 in the Faculty of Humanities, 3,090 in the Faculty of Information Technology, Mathematics, and Electrical Engineering, 2,014 in the Faculty of Natural Sciences and Technology, 1,071 in the Faculty of Medicine, and 605 in the Faculty of Architecture and Fine Art. About 3,500 bachelor and master degrees are awarded each year, and more than 5,500 participate in further education programmes.

NTNU has more than 300 cooperative or exchange agreements with 60 universities worldwide, and several international student exchange programmes. There are, at any given time, around 2,600 foreign students at the university.

Scientists at NTNU have so far been awarded four Nobel Prizes: Lars Onsager in Chemistry (1968); Ivar Giæver in Physics (1973) and Edvard Moser and May-Britt Moser in Medicine or Physiology (2014).

Source http://en.wikipedia.org/wiki/Norwegian_University_of_Science_and_ Technology and for detailed topical information: www.ntnu.edu/

NTNU's Department of Sociology and Political Science

The Department of Sociology and Political Science at NTNU (ISS) offers bachelor and master studies in sociology, political science, and sports science, as well as master studies in media, communication, and information technology (MKI) and PhD programs in sociology and political science. Studies in sociology include courses in organisation and working life, social inequality and welfare, and media. Studies in political science include courses in international and comparative politics, public policy and administration, political theory, and political behaviour. Studies in sport science include courses in sport as activity and practical area, sports sociology, and child and youth sports. MKI is a multidisciplinary programme of study combined of courses from sociology, political science, media, psychology, educational science, and information technology. All programmes of study include courses in research methods. ISS has an active research environment consisting of several research groups working on local, national, and international projects. This department cooperates with national and international partners, as well as other departments at NTNU and research institutes in Trondheim. ISS emphasizes contact and collaboration with external institutions, such as industry and commerce, the public sector, and voluntary organisations. The Department has had close collaboration with PRIO for many years. In addition to Nils Petter Gleditsch, Scott Gates and Halvard Buhaug have for many years held joint positions at the two institutions. For more information, see www.ntnu.edu/iss/

© The Author(s) 2015
N.P. Gleditsch, *Nils Petter Gleditsch: Pioneer in the Analysis of War and Peace*, SpringerBriefs on Pioneers in Science and Practice 29,
DOI 10.1007/978-3-319-03820-9

About the Author

Nils Petter Gleditsch (born 17 July 1942 in Sutton, Surrey, UK) is a Norwegian peace researcher and political scientist. He is Research Professor at the Peace Research Institute Oslo (PRIO). In 2009, Nils Petter Gleditsch was given the *Award for Outstanding Research* by the Research Council of Norway. In 1982 he was convicted (with Owen Wilkes) in Norway of a violation of the national security paragraphs of the penal code and given a suspended prison sentence. After studies in philosophy and economics Gleditsch became mag.art. in sociology at the University of Oslo. In 1966–67 he read sociology, social psychology, and international relations at the University of Michigan. Since 1964, Gleditsch has worked at the Peace Research Institute Oslo (PRIO), first as a student, later as researcher. He was Director of PRIO in 1972 and 1977–78. From 2002–08 he led the working group 'Environmental Factors of Civil War' at PRIO's Centre for the Study of Civil War, appointed as a Centre of Excellence by the Research Council of Norway. Since 1993 he has also been a part-time Professor at NTNU. Gleditsch was editor of *Journal of Peace Research* 1983–2010. He served as President for the International Studies Association (ISA) 2008–09. He is a member of the Royal Norwegian Society of Sciences and Letters (DKNVS) and the Norwegian Academy of Science and Letters (DNVA).

Among his books in English are: (coed., 1980): *Johan Galtung. A Bibliography of His Scholarly and Popular Writings 1951–80;* (with O. Wilkes, 1987): *Loran-C and Omega. A Study of the Military Importance of Radio Navigation Aid*; (co-ed. with O. Njølstad, 1990); *Arms Races—Technological and Political Dynamics*; (co-author with O. Bjerkholt; Å. Cappelen, 1994): *The Wages of Peace. Disarmament in a Small Industrialized Economy*; (co-ed. with O. Bjerkholt; Å. Cappelen; R.P. Smith; J.P. Dunne, 1996: *The Peace Dividend*; (co-ed. with L. Brock; T. Homer-Dixon; R. Perelet; E. Vlachos, 1997): *Conflict and the Environment*; (co-ed. with G. Lindgren; N. Mouhleb; S. Smit; I. de Soysa 2000):

© The Author(s) 2015
N.P. Gleditsch, *Nils Petter Gleditsch: Pioneer in the Analysis of War and Peace*, SpringerBriefs on Pioneers in Science and Practice 29,
DOI 10.1007/978-3-319-03820-9

Making Peace Pay: A Bibliography on Disarmament and Conversion; (co-ed. with P. Diehl 2001) *Environmental Conflict*; (co-ed. with G. Schneider; K. Barbieri 2003): *Globalization and Armed Conflict*; (co.ed. with G. Schneider, 2013): *Assessing the Capitalist Peace*. He has also been guest-ed. and co-ed. of special journal issues including: *European Journal of International Relations,* 1(4): 405–574; *Political Geography* 26(6): 627–735; *International Studies Review* 12(1): 1–104; *International Interactions* 36(2): 107–213; *Conflict Management and Peace Science* 28(1): 5–85; *International Interactions* 38(4): 375–569; *Journal of Peace Research* 49(1): 1–267; *International Studies Perspectives* 13(3): 211–234; *International Studies Review* 15(3): 396–419.

Website at PRIO: www.prio.org/staff/npg

Website at NTNU: www.ntnu.edu/employees/nilspg

Website on this book: http://afes-press-books.de/html/SpringerBriefs_PSP_Gleditsch.htm

About the Book

This book presents Nils Petter Gleditsch, a staff member of the Peace Research Institute of Oslo (PRIO) since 1964, a former editor of the *Journal for Peace Research* (1983–2010), a former president of the International Studies Association (2008–09) and the recipient of several academic awards as a pioneer in the scientific analysis of war and peace. This unique anthology covers major themes in his distinguished career as a peace researcher. An autobiographical, critical retrospective puts his work on conflict and peace into a broader context, while a comprehensive bibliography documents his publications over a period of 50 years. Part II documents his wide-ranging contributions on globalization, democratization and liberal peace, on international espionage, environmental security, climate change and conflict and on the decline of war and more generally of violence as a tool in conflict.

- As Editor of *Journal of Peace Research* for 27 years, former President of the International Studies Association and the recipient of several academic awards, he has a high profile in peace research and international relations
- Addresses key topics in peace research
- This book is the only one of its kind—there will be no Festschrift or autobiography

Contents

A Life in Peace Research—Bibliography—Time Differences and International Interaction—Democracy and Peace—The Treholt Case—Armed Conflict and the Environment—Double-Blind but More Transparent—The Liberal Moment Fifteen Years On—Whither the Weather? Climate Change and Conflict—The Decline of War—The Main Issues—The IPCC, Conflict, and Human Security

Websites: www.springer.com/law/environmental/book/978-3-319-03819-3 and http://afes-press-books.de/html/SpringerBriefs_PSP_Gleditsch.htm

© The Author(s) 2015

171

N.P. Gleditsch, *Nils Petter Gleditsch: Pioneer in the Analysis of War and Peace*, SpringerBriefs on Pioneers in Science and Practice 29, DOI 10.1007/978-3-319-03820-9